Studies in Computational Intelligence 426

Editor-in-Chief

Prof. Janusz Kacprzyk
Systems Research Institute
Polish Academy of Sciences
ul. Newelska 6
01-447 Warsaw
Poland
E-mail: kacprzyk@ibspan.waw.pl

For further volumes:
http://www.springer.com/series/7092

Qing Liu, Quan Bai, Stephen Giugni,
Darrell Williamson, and John Taylor (Eds.)

Data Provenance and Data Management in eScience

Editors

Qing Liu
CSIRO
Hobart, TAS
Australia

Darrell Williamson
CSIRO
Acton, ACT
Australia

Quan Bai
Auckland University of Technology
New Zealand

John Taylor
CSIRO
Acton, ACT
Australia

Stephen Giugni
CSIRO
Hobart, TAS
Australia

ISSN 1860-949X e-ISSN 1860-9503
ISBN 978-3-642-44158-5 ISBN 978-3-642-29931-5 (eBook)
DOI 10.1007/978-3-642-29931-5
Springer Heidelberg New York Dordrecht London

Softcover reprint of the hardcover 1st edition 2013

Printed on acid-free paper

Springer is part of Springer Science+Business Media (www.springer.com)

Preface

With the advanced development of computer technology, eScience allows scientific research to be carried out in highly distributed environments. The research communities involved may cross multiple disciplines, laboratories, organisations, and national boundaries. The complex nature of the interactions in the eScience infrastructure, which includes instrument, data, model, application, people and computational facilities, has identified a strong need for data management.

Data management addresses research issues in designing, building, managing, and evaluating advanced data-intensive systems and applications. In these systems and applications, a large amount of raw data and intermediate data are generated and stored. Managing data provenance, the origin of data and processes involved to generate a particular data, has also been widely recognised as a critical issue of data management system.

Data provenance enables scientists to access and query the information life cycle of a particular data product. It is essential for improving the collaborators' confidence in sharing/consuming data products, and to encourage users to participate in research collaborations. By making the data and its associated provenance information accessible for researchers, it not only helps to determine the data's value, accuracy, authorship, publication and regulation, but also allows domain scientists to work more efficiently, achieve greater exposure and protect data from loss.

This book consists of seven chapters that discuss the cutting edge data management and data provenance techniques in eScience applications. We are pleased to include work originating in Australia, Canada, Germany, New Zealand, Pakistan, the United Kingdom and the United States.

The existing provenance models (e.g. the Open Provenance Model) are designed in a technology-agnostic manner. To make the model usable in eScience, domain specialisation is required. Furthermore, most of provenance research focus on provenance capturing and representation. However, provenance data is only useful if it provides additional value to users. The three chapters of Part I present how to make provenance collected over time usable through specialisation and knowledge discovery.

In Chapter 1, Curcin, Danger and their colleagues models Randomised Clinical Trial processes by extending OPM. It ties the provenance representation closer to medical experts so that meaningful analysis could be conducted. Chapter 2 discusses how Hidden Markov Model assumptions can be applied to evaluate the workflow trust by Naseri and Ludwig. The main idea of the work is that the state of the trust of the workflow can be determined by the state of the workflow at the previous time stamp and observing the service that was executed at that time. The trust level is defined based on the transition probability and sensor probability. In Chapter 3, Aktas, Plale, Leak and Mukhi investigate the problem of unmanaged workflows from three perspectives: provenance capture, provenance representation and use. Particularly, the authors discuss how to use provenance to aid workflow construction, to reconstruct process traces and to analyse workflow traces.

Distributed provenance raises the issue of efficient reconstruction during the query time. Chapter 4 in Part II offers some query mechanisms for such situation. Chapter 5, 6 and 7 introduce some data management systems for real world case studies.

In Chapter 4, a hybrid approach to answer provenance path queries is adopted by Malik, Gehani, Tariq and Zaffar. A sketching method is proposed for querying across distributed systems and for querying within a system, transitive closure is applied. Yao, Zhang and their colleagues focus on the use of mobile and cloud techniques in data provenance. In Chapter 5, they firstly point out some current challenging issues in data provenance. A prototype mobile cloud system to solve these issues is presented and demonstrated in the bioinformatics domain. In Chapter 6, a number of data provenance/management challenging issues in radio astronomy research are addressed by Mahmoud, Ensor and their colleagues. According to authors' real experience in the Square Kilometre Array (SKA) project, a stream-computing approach has been introduced to cover some of the addressed issues. The approach is based on IBM InfoSphere Streams, and can effectively manage huge volume of data collected from large antennae array. Chapter 7 introduces a scientific data management system by Ney, Kloss and Schreiber, called DataFinder, to manage laboratory records in a computer based infrastructure. Through the usage of such a system, laboratory data can be traced and managed in a more effective and efficient way. The reliability and transparency of scientific results can also be enhanced.

Each submitted chapter was reviewed by at least 2 reviewers during 2 rounds. The editors would like to thank the chapter authors and the following external reviewers (listed in alphabetical order):

Dr Paul Groth, VU University of Amsterdam, The Netherlands
Dr Simon Miles, King's College London, UK
Dr Heiko Mueller, CSIRO, Australia
Dr Leonardo Salayandia, The University of Texas at El Paso, USA
Dr Eric Stephen, Pacific Northwest National Laboratory, USA
Dr Kerry Taylor, CSIRO, Australia

Dr Qing Zhang, CSIRO, Australia
Dr Ying Zhang, The University of New South Wales, Australia

This book would not have been possible without your contributions. Thank you all.

February 2012

Qing Liu
Quan Bai
Stephen Giugni
Darrell Williamson
John Taylor

Contents

Part I
Provenance in eScience: Representation and Use

Chapter 1
Provenance Model for Randomized Controlled Trials

Vasa Curcin, Roxana Danger, Wolfgang Kuchinke, Simon Miles, Adel Taweel, and Christian Ohmann

Abstract. This chapter proposes a provenance model for the clinical research domain, focusing on the planning and conduct of randomized controlled trials, and the subsequent analysis and reporting of results from those trials. We look at the provenance requirements for clinical research and trial management of different stakeholders (researchers, clinicians, participants, IT staff) to identify elements needed at multiple levels and stages of the process. In order to address these challenges, a provenance model is defined by extending the Open Provenance Model with domain-specific additions that tie the representation closer to the expertise of medical users, and with the ultimate aim of creating the first OPM profile for randomized controlled clinical trials. As a starting point, we used the domain information model developed at University of Dusseldorf, which conforms to the ICH Guideline for Good Clinical Practice (GCP) standard, thereby ensuring the wider applicability of our work. The application of the model is demonstrated on several examples and queries based on the integrated trial data being captured as part of the TRANSFoRm EU FP7 project.

1.1 Introduction

Within the medical domain, there is a pressing need for an extensible provenance model to enable both auditability and accountability in software. Making electronic

Vasa Curcin · Roxana Danger
Imperial College London, London SW7 2AZ
e-mail: {vasa.curcin,r.danger}@imperial.ac.uk

Wolfgang Kuchinke · Christian Ohmann
University of Dusseldorf
e-mail: {kuchinkw,Christian.Ohmann}@uni-dusseldorf.de

Simon Miles · Adel Taweel
King's College London
e-mail: {simon.miles,adel.taweel}@kcl.ac.uk

Q. Liu et al. (Eds.): Data Provenance and Data Management in eScience, SCI 426, pp. 3–33.
springerlink.com

systems provenance-aware enables users to investigate data sources and services that produced a particular output from the system, together with the individuals who instigated the requests and received those outputs. In such way, the software can be audited to assess that correct decisions were made and appropriate procedures followed.

Randomized controlled trials (RCT), also referred to as randomized clinical trials, are widely accepted as the best test of therapeutic efficacy, and as such are subject to particularly stringent quality controls to ensure legal, organizational, and scientific procedures have been adhered to. ICH Guideline for Good Clinical Practice (GCP) standard is an international ethical and scientific quality standard for the design, conduct and record of research involving humans, and it states that *All clinical trial information should be recorded, handled, and stored in a way that allows its accurate reporting, interpretation and verification.* [2]

From the computer science point of view, typical for this type of research is the necessity to eliminate the assumption of data and services being freely available. Clinical research happens in a strictly closed world, where each data exchange needs to be justified, and approved on multiple levels (patient, institution, country) which adds complexity both to the provenance of software actions involving that data, and the management of any data derived from those actions. The models already present in the domain are by and large ontologies and controlled vocabularies, defining international terminology standards, and process models specifying, often informally, the required actions during the trial. Therefore, it is essential that the provenance models link to existing standards by reusing their ontologies and provide clear mapping to their process models.

The provenance requirements for clinical research and trial management of different stakeholders (researchers, clinicians, participants, IT staff) need to be investigated to identify provenance needs at multiple levels and stages of clinical research process. Our work was done within the context of a particular clinical study, conducted within the EU TRANSFoRm project, as an exemplar to emphasise real-world provenance challenges such as privacy and security requirements arising from legal and logistical constraints. It was the bridging of data from the care context with data from registers and clinical research in TRANSFoRm that made it necessary to develop, together with data privacy and data security frameworks, a distinct data provenance framework. Further requirements within the domain are also presented with respect to traceability and subsequent provenance record analysis, including questions on task ownership, study properties, and degrees of anonymization of various parts of contributing data.

Section 1.2 introduces the clinical trial conceptual model, conforming to the GCP standard. Key aspects of the Open Provenance Model (OPM) are presented in section 1.3. Section 1.4 describes OPM-RCT, the new profile for randomised controlled trials, while examples of its usage in querying the provenance repositories are given in section 1.5. Section 1.6 summarizes the work done and lays out future directions both within TRANSFoRm and beyond.

1.2 Process Flow of Clinical Trials

Clinical trials are research studies into particular medical questions that work with human data, and follow a protocol. A clinical trial consists of three phases: (1) planning and preparation, (2) trial conduct and (3) analysis and reporting of results (see legend of diagram).

The model, as used for the OPM-RCT profile, including steps involved and the documents produced and used in each step, is based on the ICH Good Clinical Practice guideline, a standard which dates back to 1996. when it was produced by EU, US, and Japanese regulatory bodies, and has since been enshrined in law following the introduction of the EU Clinical Trials Directive 2001 and its UK implementation, UK Medicines for Human Use (Clinical Trials) Regulation 2004.

1.2.1 Trial Planning and Development

The most important preparatory step is the creation of the *trial protocol*. It describes objectives, design, methodology, statistical considerations, and the organisation of a clinical trial. The content and structure of the trial protocol is determined by the ICH GCP E6 guideline. Protocol feasibility research goes into the formulation of the protocol insuring that a suitable patient population exists and that the clinical trial conduct is viable. For the document and records management during the trial, a Trial Master File (TMF) and an Investigator Site File (ISF) have to be prepared. Both files contain relevant documents and are gradually filled with study documents during the course of the clinical trial. Trial support has to be prepared, including investigator training, development of the investigator's brochure containing efficacy and safety information, development of patient information and informed consent forms, and the development of data management plan, monitoring plan and statistical analysis plan. For medicinal product studies the logistics of drug supply has to be organised, including packaging, labelling, shipment, and accountability management. Trial quality management includes the preparation of standard operating procedures (SOPs) and audits by the sponsor. For the collection of patient data special forms, the Case Report Forms (CRF) have to be prepared. The drug supply for patient treatment has to be prepared, including the storage, distribution and management of the Investigational Medicinal Product (IMP), and the generation of a drug inventory log and temperature log. Administrative activities that accompany the trial include the negotiation of insurances, contracts and the management of trial finances. The clinical trial team has to be formed based on the responsibility split and the accompanying signature sheet. Appropriate sites have to be identified, recruited for the trial and qualified. In addition laboratories and additional service providers have to be identified, enlisted and qualified.Prior to the commencement of a trial the approval of the Competent Authority (CA) and a positive vote of an Ethics Committee (EC) must be obtained. To obtain an approval documents have to be prepared and submitted, including patient information and informed consent form, and insurance confirmation. Any clinical trial on a medicinal product also requires a clinical trial authorisation (CTA) from the CA in the EU member states in which the trial is being carried out.

1.2.2 Conduct of the Clinical Trial Process (Trial Management)

The clinical trial starts with the initiation of sites and the recruitment of patients, but only after the approval has been obtained. The data management plan is implemented and patients are screened for participation. Patients that meet the inclusion criteria are randomised and, provided the consent is given, their data is collected, cleaned by a query process and stored (data and records management); data and medical reviews are performed according to the data management plan. For data collection an infrastructure consisting of EDC system (Electronic Data Capture), web server, and a clinical database has to be set up. Every data management system used in clinical trials has to be system validated. During clinical trial conduct accompanying quality measures, such as monitoring and audits by the sponsor, have to be in place. The monitor performs initiation visits at the outset, monitoring visits during the trial and close out visits at the end of the trial. Enabling faultless safety management in clinical trials is one of the most important issues. All adverse events are recorded, however serious adverse reactions (SAR) and suspected unexpected serious adverse reactions (SUSAR) require an additional risk assessment and must be reported within a fixed period (SAE management). Serious adverse events (SAE) are adverse events that result in death or are life-threatening, require inpatient hospitalisation, or result in significant disabilities or damage, etc. Adverse Events recording has to be harmonised with SAE reporting (SAE reconciliation). Interim Safety Reports are also created during the process.

1.2.3 Trial Ending

After all patients have been recruited and all trial related procedures performed according to trial protocol, the sites are closed and the database is locked; the data collected is then sent to the sponsor or leading investigator for analysis. The trials differ in their aims: industry sponsored trials aim at marketing authorisation, while investigator initiated trials aim to improve scientific evidence about interventions and develop reports and publications. The finalisation of the trial includes the creation of the statistical report with the analysis of all results according to the statistical analysis plan, and the final trial report. The end of a clinical trial should be communicated to both the Competent Authority and the Ethics Committee. The trial ends with the archiving of all clinical study documents, with the trial master file (TMF) including the locked study database, archived by the sponsor, and the investigator site file (ISF) at the corresponding clinical centre (site).

1.2.4 Trial Metadata Analysis

The metadata of clinical studies allows the researchers, auditors, and any other authorised parties to examine the evolution of a study design for a range of purposes, from comparing multiple studies for differences in approach, via ensuring regulatory compliance, to tracing a particular investigator or participant.

Currently, the audit information about the trial, changes, and amendments is stored in individual, often paper-based documents, that are part of the Trial Master File, containing items such as notes, literature, meeting minutes, protocol draft versions, trial feasibility information etc. With respect to the conduct of a study, the process already incorporates a number of documents, e.g. data audit trail, confirmation by signature, monitoring of trial conduct, patient recruitment rate, data querying, and source data verification. Some of this information is stored in paper documents, while some is kept in the trial database.

Therefore, any analysis of the study lifecycle involves searching through a collection of paper documents, computer files, and database repositories and, by necessity requires a large number of people, from administrative personnel to clinicians and database managers. Especially in international clinical trials over different time zones, protocol amendments and changes in eCRFs may be implemented at different time points at different sites. A computerised provenance storage centralises this disparate information in a single resource that can be made accessible as needed to various user profiles, giving each an appropriate view of the data stored.

1.2.5 ICH GCP and Other Models

The OPM-RCT is based on the ICH Good Clinical Practice guideline. However, there are also other tools and models that aim to standardize the conduct of clinical trials and their data representation.

Clinical Trials Toolkit [11] was designed to help meet the requirements of the UK Medicines for Human Use (Clinical Trials) Regulation 2004, and through it the EU Clinical Trials Directive 2001. Its specification consists of three process maps covering different stages in the trial process, so called "stations." Each station has associated with it one or more resources: documents, links to other web resources, forms and others. In addition to that, each station is also classified as being either standard practice, legal requirement, or good practice, and for each of these whether it is specific to Directive's requirements, or relevant to all trials.

In the United States, Clinical Data Interchange Standards Consortium (CDISC), Health Level Seven (HL7), the National Cancer Institute, and Cancer Biomedical Informatics Grid (caBIG), have combined their individual efforts in representing trial data to develop a single standard domain model of regulated clinical research in the biomedical domain, which combines the terminologies and domains from their previous work, most notably HL7 Resource Information Model 3.0 [1]. The result is the Biomedical Research Integrated Domain Group (BRIDG) model [15] that aims to facilitate integration of disparate models within the clinical research area, integration of biomedical/clinical research with healthcare data, while being understandable to the clinical users. As such, it is not specific to any single standard.

Finally, Primary Care Research Object Model [36] aims to create a link between the reference model of clinical research, as defined by BRIDG, and the real-world design and implementation of systems to support the design, execution, analysis, and report of clinical trials in primary care research. It is based on UML methodology and designed with the view of using it for clinical trial data management systems, and other software tools. Like HL7, it is a general model that does not promote a particular standard.

1.3 Provenance

Provenance, also referred to as lineage or pedigree, refers to where something comes from or why it holds a certain property. When viewed as a technical issue with regard to some processes, such as with clinical trials, provenance typically refers to documentation of the processes that have occurred and the capture and subsequent access to that documentation, i.e. the infrastructure required to find the provenance of data produced by those processes. In comparison to features such as auditing, provenance focuses on tracking the interoperation of processes across different systems, stages and authorities, so that the full set of influences on some data output can be understood. As a research topic, provenance is currently receiving considerable attention in domains such as healthcare, banking/finance, science, journalism, and many others [27].

While many of the technical challenges around provenance are the same as for any structured data to be stored over a long term, most notably access and curation, the *model* used for provenance must be suited to its particular characteristics. Many high-level models of provenance have been proposed over recent years, including the Open Provenance Model [28], described further below. Due to this proliferation, and the many specialist representations of provenance developed in specific application domains, a W3C group has been established to produce a recommended standard model for provenance interchange [39].

1.3.1 Provenance in Healthcare

In the case of distributed medical applications, the data (electronic health records and instrument data), the workflows (procedures carried out to perform analysis on healthcare data) and the record of those workflows may be distributed among multiple heterogeneous information systems. These information systems may be under the authority of different healthcare actors such as general practitioners, hospitals, hospital departments, etc. which form disconnected islands of information. In such systems, provenance technology may take on the additional role of verifying the adherence of actors in the process to security policies.

There is a wide range of use cases which may underlie the need for provenance [26]. Users may wish to simply better interpret their data through knowing its source, or to decide whether to rely on it based on their trust for that source. As provenance documents the enactment of procedures, it can be used to analyse whether procedures are being performed efficiently, which is important in time-critical applications such as organ transplantation [22]. The involvement of private patient data means that audit and records of approval for releasing data are fundamental aspects of provenance dictated by local, national and international regulations.

Clinical trials are by nature distributed, as the sources of patient data are commonly distinct from their users. In projects such as TRANSFoRm, the interrogation of data from multiple electronic sources, at independent sites, means that an additional layer of system distribution is added. In order to execute meaningful queries regarding provenance, the provenance data needs to be interconnected and potentially integrated. To make this possible, a common model for provenance across all stages and locations of the workflow is desirable.

1.3.2 Open Provenance Model

The Open Provenance Model (OPM) [28] is a representation of the processes which have led to data being produced or transformed into a new state, and so can represent the provenance of one or more data items. Here we will summarise the relevant aspects from the OPM v1.1 specification [28].

OPM is a causal graph model of provenance, meaning that an OPM description of provenance is a graph whose edges denote causal relationships (X was caused by Y) between the occurrences denoted by the nodes. This structure allows OPM graphs to describe how multiple events led to some data being produced (serially or independently), how one piece of data was derived from another, etc. OPM classifies occurrences (nodes) into three types: *artifacts*, *processes* and *agents*. Artifacts are pieces of data of fixed value and context, e.g. one version of a document. Processes are (non-instantaneous) actions which are performed *using* artifacts to *generate* other artifacts, e.g. a random selection process uses the full set of eligible patients and generates a subset of these with which to conduct the trial. The artifacts used by a process can play different roles in the process, e.g. a document being edited versus the additions being made to the document. Agents denote the entities *controlling* process execution, potentially in different roles, such as researchers and clinicians.

The properties which artifacts, processes and agents possess can be documented by arbitrary key-value *annotations* to the nodes. Edges can also have annotations to provide further information on *how* one occurrence caused another. OPM allows for multiple levels of granularity of description of the same set of past events, with each description being a separate *account* of what occurred. Specifically, what is represented as a single process in one account can be *refined* in another account to describe how that process is decomposed into multiple sub-processes with intermediate data being generated and used.

In the following sections, we show illustrative OPM graphs, which use the following conventions: *(i)* circles are artifacts, rectangles are processes, and octagons are agents; *(ii)* an edge from an artifact to a process is of type `wasGeneratedBy`, from process to artifact is `used`, from process to process is `wasTriggeredBy`, from artifact to artifact is `wasDerivedFrom`, from process to agent is `wasControlledBy`; *(iii)* we omit these types from the figures for brevity, while if a role is specified, where the role is shown in brackets as in `(R)` — if not specified, it has the value `undefined`.

1.3.3 Profiles

A generic provenance model such as OPM provides the high-level concepts required to generate and query provenance data across interconnected systems. When used in practice, it must be augmented with application-specific terminology and data, i.e. vocabulary for expressing particular *kinds* of artifact, process, agent, etc.

In many domains, such terminology requirements are not specific to a single deployment. In such cases, rather than each separate team solving a similar problem determining their own vocabulary to use in conjunction with the generic provenance model, it is preferable that there is some agreed community standard extension to the generic model. In OPM as elsewhere, this is called a *profile*. By defining a profile for provenance in clinical trial research, we aim to allow far greater interoperability between systems involved in supporting such trials. Without a profile, it is doubtful that deep interoperation of systems would be useful, because critical questions regarding the provenance of the data they use and produce would be unanswerable.

An OPM profile is intended to define a specialisation of OPM, and it consists of the following elements.

- A unique global identifier for the profile
- A controlled vocabulary
- Guidance on how to express OPM graphs in the domain of this profile
- Expansion rules stating how concepts introduced in the profile translate into generic OPM graph structures
- Profile-specific syntax to be used in serialisation of OPM graphs

As with OPM itself, a profile can be defined independently of how data using that profile is serialised (excepting the optional serialisation syntax mentioned above). OPM does not specify how data using the profiles must be serialised, and there are a variety of formats available, including two OWL[1] ontologies (OPMO [29] and OPMV [45]) with different levels of expressivity, and RDF serialisations.

[1] OWL is the acronym for Web Ontology Language [24]

1.4 OPM RCT Profile Proposal

This section describes the OPM profile proposal for modelling Randomized Controlled Trial processes. Its URI identifier is defined as `http://www.lesc.imperial.ac.uk/rcto/rctpo`, and in the rest of the chapter *rctpo:* will be used as the shortened prefix for the profile. We now proceed to present the controlled vocabulary for the profile, its expansion rules, and give several examples.

1.4.1 RCT Controlled Vocabulary

The OPM-RCT proposal requires an underlying ontology to capture the RCT domain concepts in the provenance model. We do this by integrating several popular ontologies and controlled vocabularies, and extending them with missing concepts and relations to define the Randomized Controlled Trials Ontology (RCTO) that serves as the back-end to OPM-RCT.

RCT and RCT-related concepts that appear in the ICH GCP guideline can be found in several ontologies.

- Ontology for Biomedical Investigations[2] [12] contains a formal description of the protocols, instrumentation, material, data used, results generated, and analysis performed during biomedical research.
- Biomedical Research Integrated Domain Group (BRIDG) Model[3] [44] contains the common concepts in regulated clinical research protocols. The current version describes in detail the protocol, the study conduct and adverse event data, as well as a hierarchy of all concepts involved in clinical trial studies.
- Ontology of Clinical Research[4] [35] contains a formal description of human studies and the studies elements. It does not contain details of study conduct yet, but includes a hierarchy of study types and status, as well as a set of administrative processes during the planning phase that are not described in any other controlled vocabulary.
- Epoch Clinical Trial Ontology[5] [34] is composed of seven ontologies (Clinical Trial, Assay, ConstrainExpression, Labware, Measurement, Organization and Protocol) resulting in a detailed formal description of trial protocols and the experimental conditions of assays.

---

[2] `http://purl.obolibrary.org/obo/obi`, prefix: *obi:*.

[3] `http://www.bridgmodel.org`, used prefix: *bridg:*.

[4] `http://purl.org/net/OCRe/HSDB_OCRe.owl`, used prefix: *ocre:*.

[5] `http://epoch.stanford.edu/ClinicalTrialOntology.owl`, used prefix: *epoch:*.

- Adverse Event Ontology (AEO)[6] [17] was designed with the aim of standardising adverse event annotation, integrating various adverse event data, and supporting computer-assisted reasoning.

All of the above have been defined using OWL, except for BRIDG, which is based on UML[7].

These ontologies and controlled vocabularies are complementary in some concepts, and overlapping in others, therefore, an ontological alignment is needed in order to provide interoperability of the systems using them. Also, although some types of entities, such as documents, electronic clinical resources, roles and processes are already described in these ontologies, complete and rigorous conceptual hierarchies for them are completely lacking.

To overcome these challenges a new ontology is introduced, *Randomized Controlled Trial Ontology*, at `http://www.lesc.imperial.ac.uk/rcto/rcto` and with the shortened prefix *rcto:*. RCTO is an extension of the OBI ontology, with the following features:

- The ontology alignment is included in the concepts of *rcto*.
- A hierarchy is introduced for Clinical Trial Documents, an extension of the *obi:Document* concept, by characterizing each of the documents, and the relationships between them. The full description, using Description Logic statements, is in Table 1.1.
- A hierarchy is introduced for Electronic Clinical Resources, which is an extension of the *obi:information_content_entity* concept, as described in Table 1.2.
- The hierarchy of *obi:roles* has been refined in accordance with the hierarchy in BRIDG model.
- As the above ontologies are all missing some fundamental process concepts (namely Trial protocol definition, Trial master file and essential document preparation, monitoring plan preparation, statistical analysis preparation, CRF creation, and others), they have been introduced to the new process hierarchy.

Note that *rcto:* specifies only the clinical trial concepts derived from the above proposed extensions, as opposed to *rctpo:*, which specifies OPM-RCT concepts.

The OPM-RCT vocabulary proposal is described in subsections 1.4.1.1 - 1.4.1.5 and a summary of a formal description of some representative entities definition are given in subsection 1.4.1.6. The elements of the vocabulary are annotated onto the OPM artifact, agent and process entities through OPM Annotations framework and the *rdf:type* property, effectively declaring subtypes.

[6] `http://bioportal.bioontology.org/ontologies/1489`, used prefix: *aeo:*.

[7] UML is the acronym for Unified Modeling Language [10]

Table 1.1 Proposal extension for Clinical Trial Document hierarchy. *rcto* is used as the prefix of the proposed extensions.

Class	Definition	Notes
rcto:authDocument	$\sqsubseteq$ *obi:edited_document* $\sqcap$ $=1$ *dc:creator* $\sqcap$ ≥ 1 *dc:contributor* $\sqcap$ $=1$ *dc:available* $\sqcap$ $=1$ *dc:dateAccepted* $\sqcap$ $=1$ *dc:modified*	An authorship document is an *obi:edited_document* which defines its creator, the submission and acceptance date, and update method used.
rcto:TrialProtocolDocument	$\sqsubseteq$ *rcto:authDocument* $\sqcap$ $\sqsubseteq$ *bridg:StudyProtocolDocument* $\sqcap$ ≤ 1 *rcto:hasProtocol.(obi:protocol* $\sqcup$ *bridg:StudyProtocol* $\sqcup$ *epoch:Protocol* $\sqcup$ *ocre:StudyProtocol)*	A trial protocol document is a subconcept of both *rcto:authDocument* and *bridg:StudyProtocolDocument*, which can be associated to a *Protocol* object as defined in *obi*, *bridg*, *epoch* or *ocre* ontology.
rcto:FeasibilityAnalysisDocument, *rcto:TrialMasterFile*, *rcto:InvestigatorSiteFile*, *rcto:StudySiteFileDescription*, *rcto:LabsFileDescription*, *rcto:TeamFileDescription*, *rcto:MonitoringPlan*, *rcto:ClinicalReportForm*, *rcto:QuestionnaireForm*, *rcto:StatisticalAnalysisDescription*	$\sqsubseteq$ *rcto:authDocument*	Each of these concepts represents a specific document type as described in the Clinical Trial specifications, and may be expanded further with a detailed set of distinctive features.
rcto:SuppliesFileDescription	$\sqsubseteq$ *rcto:authDocument* $\sqcap$ $\sqsubseteq$ *epoch:FacilitiesPlan*	A supplies file document is a subconcept of both *rcto:authDocument* and the more specific *epoch:FacilitiesPlan* object which describe the supplies facilities of the trial.

Table 1.1 *(continued)*

Class	Definition	Notes
rcto:AdministrativeDocument	$\sqsubset$ *rcto:authDocument* $\sqcap$ ≤ 1 *obi:has_specified_output.ocre:Agreement_to_perform_study_activity*	This concept represents administrative documents needed for the study performance. Note that 'Agreement_to_perform_study_activity' is a process defined in *ocre* ontology.
rcto:ApprovedTrialDocument	$\sqsubset$ *rcto:authDocument* $\sqcap$ ≤ 1 *obi:has_specified_output.ocre:EthicsApprovalProcess*	
rcto:ClinicalReport, *rcto:DataAnalysisReport*	$\sqsubset$ *rcto:authDocument* $\sqcap$ $\sqsubset$ *obi:report*	
rcto:TrialConclusionsReport	$\sqsubset$ *rcto:authDocument* $\sqcap$ $\sqsubset$ *obi:report*	
rcto:FilledQuestionnaire, *rcto:InformedConsent*, *rcto:pCRF*, *rcto:InvestigatorWorksheet*	$\sqsubset$ *obi:information_content_entity* $\sqcap$ ≤ 1 *rcto:hasImage.obi:image*	These concepts represent documents for which a paper format should exists.

Table 1.2 Proposal extension for Electronical Resources hierarchy. *rcto:* is used as the prefix of the proposed extensions.

Class(es)	Definition	Notes
rcto:ElectronicResource	$\sqsubseteq$ *obi:information_content_entity* $\sqcap$ $= 1$ *dc:creator* $\sqcap \geq 1$ *dc:contributor* $\sqcap$ $= 1$ *dc:accrualMethod* $\sqcap = 1$ *dc:available* $\sqcap$ $= 1$ *dc:modified* $\sqcap = 1$ *rcto:accession.xsd:string*	An electronic resource knows its creator, updating method, available, last modified date, and a URI to the actual electronic resource.
rcto:eMedicalRecord	$\sqsubseteq$ *rcto:ElectronicResource* $\sqcap$ $\sqsubseteq$ *obi:eMedical_Record*	
rcto:eCaseReportForm	$\sqsubseteq$ *rcto:ElectronicResource* $\sqcap$ $\sqsubseteq$ *obi:electronic_case_report_form*	
rcto:eSource	$\sqsubseteq$ *rcto:ElectronicResource* $\sqcap$ $\sqsubseteq$ *obi:eSource_document*	
rcto:DataBase	$\sqsubseteq$ *rcto:ElectronicResource* $\sqcap$ $= 1$ *rcto:hasConnectionSettings.rcto:ConnectionSettings*	
rcto:LinkedData	$\sqsubseteq$ *rcto:ElectronicResource* $\sqcap$ $= 1$ *rcto:hasConnectionSettings.rcto:ConnectionSettings*	
rcto:System	$\sqsubseteq$ *rcto:ElectronicResource*	The *rcto:accession* should contain the access point to the system in question, if public, or an informative URI, if not.

1.4.1.1 Artifacts

Artifacts describe an instantaneous state of an entity in the application. In OPM-RCT applications, we introduce three new types of artifacts:

- **Document** (*rctpo:DocumentArtifact*): A type of artifact to characterize a clinical trial document.
- **Observation Result** (*rctpo:ObservationResultArtifact*): A type of artifact to characterize an observation (clinical intervention) result which was performed during the trial.
- **Electronic Resource** (*rctpo:ElectronicResourceArtifact*): A type of artifact to characterize electronic resource used/accessed during a trial.

In provenance graphs, these three types represent the exact state of documents, data observation results, and electronic resources in relation to individual process instances, e.g. exact version of a trial document that resulted from a completed approval process.

1.4.1.2 Agents

Agents are entities which control the processes of an application. In RCT application an agent can be:

- **Person** (*rctpo:PersonAgent*): Physical actor in the process who can perform a number of roles (see below) depending on the exact function with relation to the task.
- **System** (*rctpo:SystemAgent*): System is a program or computational module for data control and processing.

The distinction between the two is introduced to clearly separate automated steps, performed by a computer system, from those that require human intervention, which we foresee to be important for auditing purposes.

1.4.1.3 Processes

Processes are entities characterizing the activities of an application. As explained in Section 1.2, an RCT trial is divided in three phases: *planning*, in which the trial is planned; *trial conduct*, in which, after the trial approval, all planned events to validate the trial hypothesis are performed; and *analysis*, in which data collected during the trial conduct phase are analysed and the conclusions of the trial are given.

Processes in the RCT application can be of different subtypes, according to their general goal:

- **Documentation** (*rctpo:DocumentationProcess*): a process which has to produce, at the very least, a document describing the results of its processing. That is, all OPM graphs containing a *Document preparation* process also contain a dependence with a *rctpo:DocumentArtifact*.

- **Support preparation** (*rctpo:SupportPreparationProcess*): a process to describe how to select and prepare staff, materials and sponsors of the trial.
- **Intervention** (*rctpo:InterventionProcess*): a process to describe the applied patient interventions (observations, in the *bridg* terms) during the trial.
- **Data publication** (*rctpo:DataPublicationProcess*): a process which describes how data are introduced and published.
- **Data analysis** (*rctpo:DataAnalysisProcess*): a process which describes how data can be accessed and analysed; as a result, at least one document, written by a study researcher and describing the conclusions of the analysis, should be returned.

In the following, we enumerate a set of more specific RCT processes, organised according to the phase in which they are involved:

I. Planning phase
   This phase is characterized by a set of *documentation* processes:

   - Trial master file and essential document preparation
   - Monitoring plan development
   - Patient information and inform consent (*rctpo:PatientInformationAnd Consent Process*)
   - eCRF development, validation and data management plan development
   - Statistical analysis plan development
   - Administrative documents preparation
   - Quality management and SOPs preparation

   There are other processes related with *support preparation*:

   - Trial team formation
   - Study sites identification
   - Patient randomization preparation
   - Recruitment phase: informed consent collection
   - Trial personal support preparation
   - Drugs and other interventions supplies preparation
   - Labs identification, qualification and preparation

   The last three processes above are of *rcto:DocumentationProcess* type, as they have to guarantee a documented output describing the selected entities for the trial execution. In particular, a *Trial personal support preparation* process contains a dependence with an *rcto:InvestigatorSiteFile* artifact; *Drugs and other interventions supplies preparation*, with the *rcto:SuppliesFileDescription* artifact; and *Labs identification, qualification and preparation*, with the *rcto:LabsFileDescription* artifact.

   Figure 1.1 represents the planning phase of an example RCT. The *trial planning* process triggered four processes. The *Trial master file and essential document preparation, Labs identification, qualification and preparation* and *Monitoring plan development* processes generated the *TrialMasterFile*, the LabsFileDescription and the *MonitoringPlan* documents, respectively. *Trial master file and essential document preparation* was guided by the Guideline for Good Clinical Practice [2] document (the `specifiedBy` dependence, an specialization of the `used` dependence, is explained in Section 1.4.1.4). *Recruitment phase* was also executed as consequence of the beginning of the planning phase, and controlled by an investigator, who explained the study to the patient. Following that, the investigator answered patient's questions, and the patient filled in the informed consent. The *Trial protocol* document for the study was then generated as the final result of the *Trial planning* process. Note that the `derivedFrom` relations were added from it to all previous generated documents.

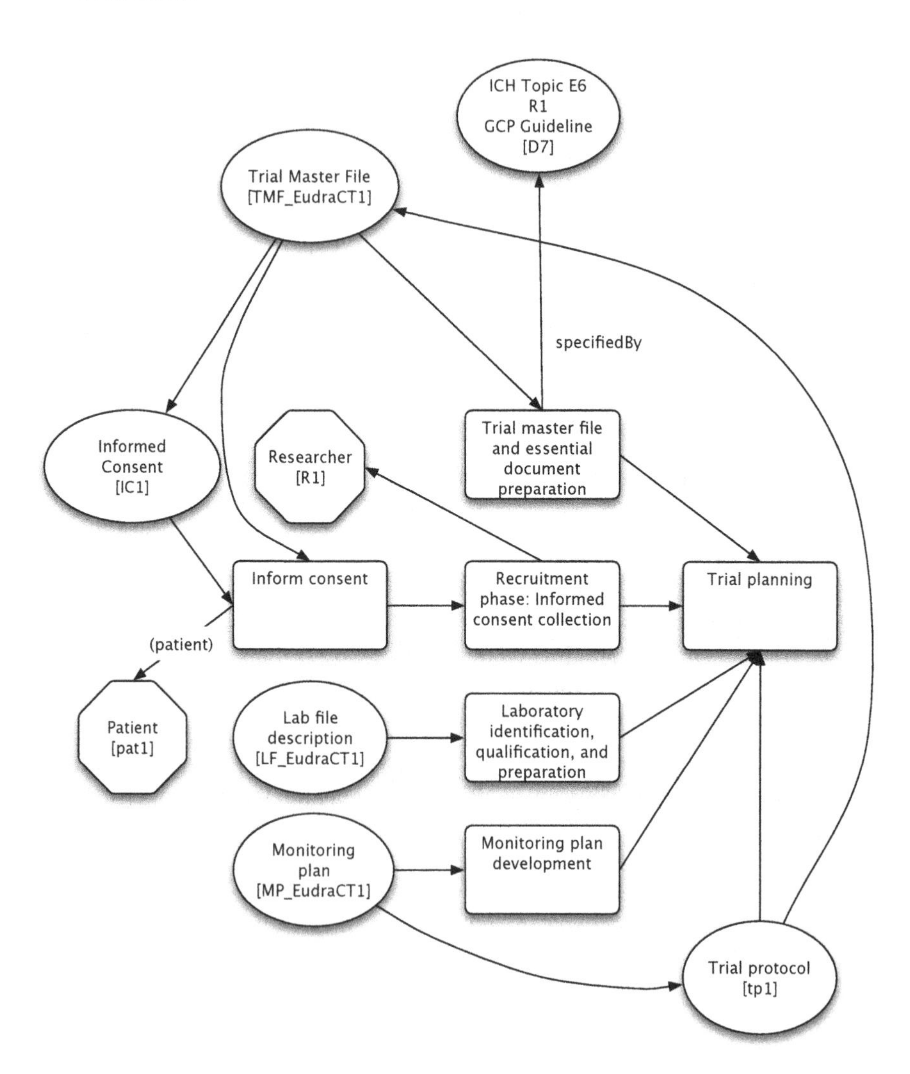

Fig. 1.1 OPM graph during the planning phase of a RCT.

II. Trial conduct phase

This phase is characterized by two major groups of processes. The *Data publication* ones include:

- Randomize selection. All patients with an informed consent are inserted into the system to perform a random selection
- Patient registering. (*rctpo:PatientRegisteringProcess*) All selected patients are introduced into a Clinical Trial Data Management (CTDM) system (that is, the execution of the *CRF creation* process for example EDC). In order to maintain their trial data and the Trial Master File and the Investigator Site File documents are adequately updated.
- CRF creation. (*rctpo:CRFCreationProcess*) A CRF is associated with each patient in a CTDM system. It takes the eMR of a patient and fill out the eCRF of the current health status of the patient.
- Patient data update. (*rctpo:PatientDataUpdateProcess*) This process modifies the patient data under the control of a system and/or person agent and the previous stored patient data, and belongs to one of the following subtypes:
  - Paper based data update; a data manager annotation based on paper notifications from the investigator or study nurse.
  - Obtaining data from worksheet; investigator notes in worksheet.
  - Electronic update data; patient data are updated automatically from a quality of life questionnaire, an intervention result, or amendments over the eCHR done by the investigator.

The *Data analysis* processes in this phase include:

- Data queries and validation. Data managers can query the trial patient database and validate some data, which can trigger a process for *Data and medical intervention reviews*,
- Data and medical intervention reviews. Investigators can review the data from interventions and other data acquisition sources in order to add detail.
- Statistical Validation. At the end of the trial, biostatisticians can query the complete database for remaining discrepancies or missing data.

Finally, the remaining processes of this phase are:

- Quality-of-life questionnaire application. Patient fills out a questionnaire about his/her current health state.
- Patient intervention. Patient is subjected to a specific planned procedure.
- Adverse event control An adverse event is controlled and, consequently, the planned procedure for such case is triggered.
- Monitor control. Monitors review documents and data for completeness and correctness (source data validation).
- Close site. Closure of a study site is performed when all data have been updated, validated and no inconsistencies have been observed.

III. Trial analysis phase
The data analysis starts after all the study sites have been closed. Two processes are present here, both of which are subtypes of both *rctpo:DocumentationProcess* and *rctpo:DataAnalysisProcess*:

- Analyse data. (*rctpo:TrialDataAnalysisProcess*) All planned analyses are performed over the collected trial data, according to the Statistical Analysis Plan, and a document is created with all the findings. The example can be found in Section 1.4.3.
- Finalization trial. (*rctpo:FinalizationTrialProcess*) The conclusions of the trial are presented in a document.

The last process in an RCT is the *Archiving trial* process, in which the trial is considered as finished.

1.4.1.4 Dependences

These are relations between different entities in the graph. In RCT application two new causal dependences are needed:

- `wasSpecifiedBy` is a specialisation of `used` establishes the relation between a process and the document artifact which proscribes how to conduct the process. For example, in Figure 1.1 it is used to describe that the Trial Master File and other essential document preparation process was constructed according to the Good Clinical Practices document.
- `laterGenerationThan`, a specialisation of `wasDerivedFrom` denotes the relation between two artifacts, in which the former is an updated version of the latter. It is used to trace the changes in the CRF of a patient, or any trial documents following update actions during the trial.

Neither of these dependences is necessarily restricted to the RCT domain. `laterGenerationThan` was also used in [25], applied to versioning of documents represented with Dublin Core terms.

1.4.1.5 Roles and Accounts

Roles in OPM-CRT serve to further specify the part that entities play in relations, for example that an agent is not only controlling a process, but is doing so as a study coordinator, or that a piece of software controlled the task in the role of a recruitment service, or that an artifact is used as input data to the process. The following roles have been identified:

- **Controlled-by (persons)**: Study coordinator, Investigator, Study nurse, Statistician, Biometrician, Data manager, eCRF designer, Head of eCRF data management, Sponsor, Patient, Clinician.
- **Controlled-by (computer systems)**: Randomization service, Recruitment service, Monitor service, eCRF based data collection, Clinical Trial Data Management system.

- **Used-by**: Input.

The OPM RCT profile also introduces several accounts, each of which is associated with a certain level of abstraction in displaying the graphs, i.e. visibility restrictions on artifact types and processes. The profiles mirror the intended user types of the system, and the segments of provenance data space that will be relevant and accessible to them. These will include study coordinators, clinicians, investigators, database administrators and data managers. The exact granularity of each of these is still under development.

1.4.1.6 Formal Definition Examples of the Proposed RCT Profile Vocabulary

In Table 1.3 some representative examples of each entity type in the proposed profile are formally defined using description logic statements.

1.4.2 Profile Expansion Rules

The expansion rules are used to introduce new relations between the entities in the OPM model. Three derived dependences are used in OPM-RCT.

[Artifact `wasDefinedBy` Agent] An artifact, a, `wasDefinedBy` an agent ag, written as: $a \to^* ag$ if the following chain of dependences can be constructed: a `wasGeneratedBy` p `wasControlledBy` ag, that is, the agent ag controlled a process which generated the artifact a.

If the artifact is a document, the `wasReviewedBy` relation can be used. It is defined as a specialization of `wasDefinedBy`.

[Artifact `wasReviewedBy` Agent] This is a specialization of the `wasDefinedBy` dependence that is created if the object is a Document artifact. A document artifact, da, `wasReviewedBy` an agent ag, written as: $da \to^* ag$ if the following chain of dependencies can be constructed: da `wasGeneratedBy` p `wasControlledBy` ag, that is, the agent ag controlled a process which generated the Document artifact da.

[Agent `wasOverseenBy` Agent] An agent, ag', `wasOverseenBy` another agent, ag, written as: $ag' \to^* ag$ if the following chains of dependences can be constructed:

1. p `wasControlledBy[study coordinator]` ag,
2. $p' \to^* p$,
3. p' `wasControlledBy` ag',

In this way, it is established who was the person overseeing the actions of some agent in a study task.

Table 1.3 Entities in the OPM-RCT profile.

Class(es)	Definition
	Artifacts
rctpo:DocumentArtifact	$\sqsubseteq$ *obi:information_content_entity* $\sqcap \sqsubseteq$ *opmv:Artifact*
rctpo:ObservationResultArtifact	$\sqsubseteq$ *bridg:PerformedObservationResult* $\sqcap \sqsubseteq$ *opmv:Artifact*
rctpo:ElectronicResourceArtifact	$\sqsubseteq$ *rcto:ElectronicResource* $\sqcap \sqsubseteq$ *opmv:Artifact*
	Agents
rctpo:PersonAgent	$\sqsubseteq$ *bridg:Person* $\sqcap \sqsubseteq$ *opmv:Agent*
rctpo:SystemAgent	$\sqsubseteq$ *rcto:System* $\sqcap \sqsubseteq$ *opmv:Agent*
	Processes
rctpo:DocumentationProcess	$\sqsubseteq$ *opmv:Process* $\sqcap$ $\sqsubseteq$ *obi:documenting* $\sqcap$ ≥ 1 *opmo:effectInverse.*($\exists$ *opmo:effect.rctpo:DocumentArtifact*)
rctpo:DataPublicationProcess	$\sqsubseteq$ *opmv:Process* $\sqcap$ ≥ 1 *opmo:effectInverse.*($\exists$ *opmo:effect.obi:content_entity_information*)
rctpo:DataAnalysisProcess	$\sqsubseteq$ *rctpo:DocumentationProcess* $\sqcap$ ≥ 1 *opmo:effectInverse.*($\exists$ *opmo:effect.rctpo:DataAcquisitionProcess*) $\sqcap$ $= 1$ *opmo:effectInverse.*($\exists$ *opmo:cause.*(*rctpo:PersonAgent* $\sqcap$ *bridg:StudyResearcher*))

Table 1.3 (*continued*)

rctpo:DataAdquisitionProcess	⊑ *opmv:Process* ⊓ ⊑ *obi:data_acquisition*
rctpo:TMFandEssentialDocsPrepProcess	⊑ *rctpo:DocumentationProcess* ⊓ = 1 *opmo:effectInverse.*(∃ *opmo:effect.rctpo:TrialMasterFile)* ⊓ = 1 *opmo:effectInverse.*(∃ *opmo:cause.(rctpo:PersonAgent* ⊓ *bridg:StudyCoordinator))*
rctpo:MonitoringPlanProcess	⊑ *rctpo:DocumentationProcess* ⊓ = 1 *opmo:effectInverse.*(∃ *opmo:effect.rctpo:MonitoringPlan)* ⊓ = 1*opmo:effectInverse.*(∃ *opmo:cause.(rctpo:PersonAgent* ⊓ *bridg:StudyCoordinator))*
rctpo:PatientInformationAndConsentProcess	⊑ *rctpo:DocumentationProcess* ⊓ = 1 *opmo:effectInverse.*(∃ *opmv:Process* ⊓ *obi:informing_subject_of_study_arm)* ⊓ = 1 *opmo:effectInverse.*(∃ *opmo:effect.(opmv:Process* ⊓ *obi:informed_consent_process))* ⊓ = 1 *opmo:effectInverse.*(∃ *opmo:effect.(opmv:Artifact* ⊓ *rcto:InformedConsent))*
rctpo:PatientRegisteringProcess	⊑ *rctpo:DataPublicationProcess* ⊓ ⊑ *obi:human_subject_enrollment* ⊓ = 1 *opmo:effectInverse.*(∃ *opmo:effect.rctpo:pCRFcreation)* ⊓ = 1 *opmo:effectInverse.*(∃ *opmo:effect.rcto:TrialMasterFile)* ⊓ = 1 *opmo:effectInverse.*(∃ *opmo:effect.rcto:InvestigatorSiteFile)*

Table 1.3 *(continued)*

rctpo:CRFCreationProcess	$\sqsubseteq$ *rctpo:DataPublicationProcess* $\sqcap$ $=1$ *opmo:effectInverse.*($\exists$ *opmo:cause.rctpo:SystemAgent*) $\sqcap$ $=1$ *opmo:effectInverse.*($\exists$ *opmo:cause.rcto:eMR*) $\sqcap$ $<= 1$ *opmo:effectInverse.*($\exists$ *opmo:cause.rcto:eCRF*)
rctpo:PatientDataUpdateProcess	$\sqsubseteq$ *rctpo:DocumentationProcess* $\sqcap$ $=1$ *opmo:effectInverse.*($\exists$ *opmo:cause.rctpo:SystemAgent*) $\sqcap$ $=1$ *opmo:effectInverse.*($\exists$ *opmo:cause.rcto:eCRF*) $\sqcap$ $=1$ *opmo:effectInverse.*($\exists$ *ompo:effect.rcto:eCRF*)
TrialDataAnalysisProcess	$\sqsubseteq$ *rctpo:DocumentationProcess* $\sqcap$ $\sqsubseteq$ *rctpo:DataAnalysisProcess* $\sqcap$ $=1$ *opmo:effectInverse.*($\exists$ *opmo:effect.DataAnalysisReport*)
FinalizationTrialProcess	$\sqsubseteq$ *rctpo:DocumentationProcess* $\sqcap$ $\sqsubseteq$ *rctpo:DataAnalysisProcess* $\sqcap$ $=1$ *opmo:effectInverse.*($\exists$ *opmo:cause.DataAnalysisReport*) $\sqcap$ $=1$ *effectInverse.*($\exists$ *opmo:effect.TrialConclusionsReport*)

1.4.3 Examples

To illustrate the usage of OPM-RCT, in this section we provide two examples of provenance graphs tracing the execution of tasks from TRANSFoRm project. Entity types are denoted by labels, while the identifiers for individual entities are written in angular brackets below.

In Figure 1.2, a trace of the analysis phase of the RCT is shown, omitting tasks preceding it and other documents. The initialization step was performed by the study coordinator `coord1`, as represented by the `wasControlledBy` relation between the agent and the process, with the agent having the role *study coordinator*. The next step, connected by `wasTrigerredBy` relation to the previous, was the statistical data analysis, done by the statistician `stat1`, and which resulted in the data analysis report, which `wasGeneratedBy` the process, and was used to derive the final trial conclusions report in the task that finalized the trial. Once this had completed, the trial was archived for future reference.

Applying the new expansion rules produces two new relations that can be added to the graph: a) statistical report `wasReviewedBy` the statistician `stat1`, and, b) `stat1` `wasOverseenBy` the study coordinator `coord1`.

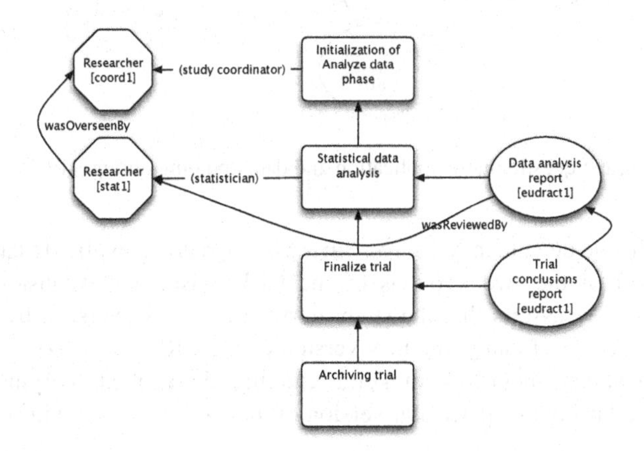

Fig. 1.2 OPM graph for the analysis phase of a RCT.

The second example, shown in Figure 1.3 involves the creation of a *quality-of-life questionnaire*.

The coordinator `coord1` controlled a process of patient data update, which was performed on the CRF entity using an eCRF software, through the *electronic data collection update* process. The data for the update was obtained from the questionnaire produced in a *quality of life questionnaire application* process, by a patient `pat1` and a study nurse, and `wasSpecifiedBy` the trial protocol `EudraCTTP1`. A new updated version of the CRF was produced and used to generate an updated version of the patient CRF study database.

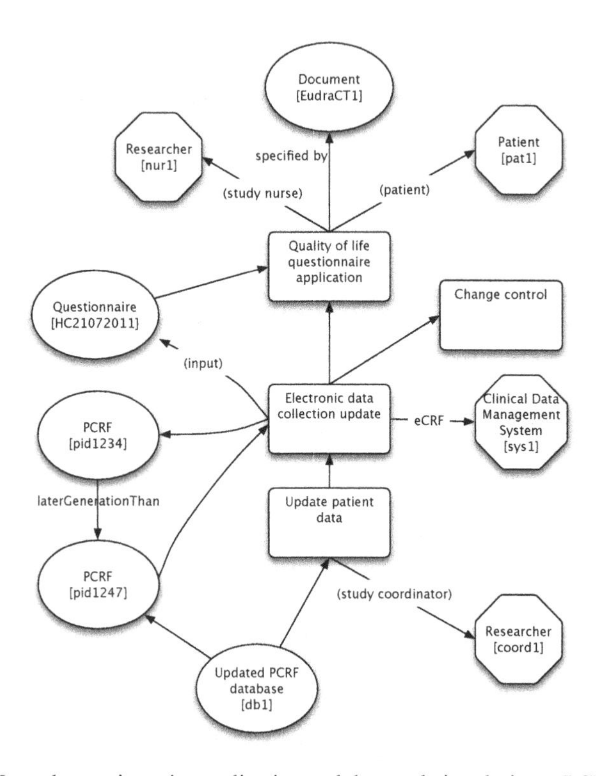

Fig. 1.3 OPM graph questionnaire application and data updating during a RCT.

The new profile dependency `laterGenerationThan` explicitly captures the change control of the two versions of the CRF. Also, the expansion rule for `wasOverseenBy` asserts that the study nurse `nur1` was overseen by the study coordinator `coord1`. Finally, the new version of the CRF `wasDefinedBy` the Clinical Data Management System `sys1`, enabling the explicit querying of documents produced using this particular version of the CDMS software in some future audit.

1.5 Storage and Analysis of Provenance Data

As presented above, OPM-RCT graphs are implemented as RDFs containing OPM-RCT profile instances, which could be very large in size, and as a requirement of a RCT application, they must be made persistent. Therefore, the OPM-RCT graphs need to be managed using one of the available RDF stores which typically use a RDBMS to manage the RDF data and support SPARQL [31], the W3C recommendation as Query Language for RDF data, or a similar semantic query language.

Jena [43, 42, 18], Sesame [7, 8], Mulgara [41], OpenLinkVirtuoso [14, 19], AllegroGraph [20, 21], BigOWLIM [5], Garlik 4store [16, 33], BigData[38], Semantic Oracle (11g)[32, 30], Intellidimension Semantic Platform[8] and RDFprov [9] are examples of this type of RDF repositories.

Amongst them, only RDFProv has been designed to also serve as a provenance metadata management system and is compliant with OPM (v1.1). It defines two types of relational database schema suitable for RDF storage, with algorithms for data insertion and SPARQL-to-SQL translation, considering the specific features of provenance data.

Non-semantic approaches for provenance data storage and querying have been described in literature, but most of them provide only export functionality from internal workflow format to OPM in XML or RDF such as VisTrails [13], eBioFlow [40], and Kepler [3]. Taverna [37], OPMProv [23] and PLIER [4], use a relational database at the back end, and reasoning and querying is expressed directly in SQL, without any inference engine present. This strategy is valid for general domains in which OMP graph entities are direct instances of the OPM ontology and not more specific objects. In knowledge domains, like RCT, where a hierarchical classification of the OPM entities is needed, inference engines increase the detail of the metadata analysis and SQL alone may not suffice.

Another solution, incorporating the best of both worlds, consists of virtualizing the provenance database as an RDF schema, using an engine such as D2RQ [6], which contains a declarative mapping between relational database schemata and OWL/RDFS ontologies. In this way, simple queries can be efficiently answered using SQL, while more complex queries can utilize SPARQL over the RDF views of the provenance data in the OPM database. Examples of using this strategy for querying provenance data in the RCT domain are given below.

The current prototype uses a variant of the database schema in [23], extended by the table *Graph(OPMGraphId, OTimeLower, OTimeUpper)*, which stores the graphs and the timestamp of their executions.

Example 1.1. The SQL queries are convenient for simple structural relationships between task instances that do not require relation roles to be considered. This query lists all the processes of executions completed on 01/08/2011 that were the starting (root) tasks.

```
1 SELECT ProvenanceDB.Process.ProcessId
2 FROM   ProvenanceDB.Process, ProvenanceDB.OPMGraph, ProvenanceDB.ExplicitWasTriggeredBy,
3 WHERE  ProvenanceDB.Process.OPMGraphId = ProvenanceDB.OPMGraph.OPMGraphId AND
4        ProvenanceDB.OPMGraph.OTimeUpper BETWEEN '01/08/2011' AND '02/08/2011' AND
5        NOT ProvenanceDB.ExplicitWasTriggeredBy.EffectProcessId = ProvenanceDB.Process.ProcessId
```

[8] http://www.intellidimension.com/products/semantics-platform/

Example 1.2. For detailed, domain-specific questions, such as the actions of various study personnel to do with certain actions, SPARQL allows us to use the ontological information defined. This query is used to find all the study nursers who have checked a screening marked as erroneous by the monitor at a specified study site.

```
1  PREFIX rdfs: <http://www.w3.org/2000/01/rdf-schema#>
2  PREFIX opmo: <http://openprovenance.org/model/opmo>
3  PREFIX rctpo: <http://www.lesc.imperial.ac.uk/rcto/rctpo>
4  SELECT ?studyNurse
5  WHERE  {
6       ?graph  rdfs:subClassOf opmo:OPMGraph .
7       ?graph  rctpo:executedIn s .
8       ?graph  opmo:hasProcess ?screenProc .
9       ?screenProc rdfs:subClassOf rctpo:ScreenProcess .
10      ?screenProc rctpo:hasStatus 'Error' .
11      ?graph  opmo:hasDependence ?dep .
12      ?dep rdfs:subClassOf opmo:WasControlledBy .
13      ?dep opmo:effect ?screenProc .
14      ?dep opmo:cause ?studyNurse .
15      ?dep opmo:Role 'Study Nurse' .
16 }
```

Example 1.3. Similarly, semantic searches are used to track the provenance of a particular entity, whether physical or electronic. This listing includes all versions of a specified pCRF, together with the date and the people involved in each modification, given the patient identifier.

```
1  PREFIX rdfs: <http://www.w3.org/2000/01/rdf-schema#>
2  PREFIX opmo: <http://openprovenance.org/model/opmo>
3  PREFIX obi: <http://purl.obolibrary.org/obo/obi>
4  PREFIX rcto: <http://www.lesc.imperial.ac.uk/rcto/rcto>
5  SELECT ?a, ?t, al
6  WHERE  {
7       ?graph  rdfs:subClassOf opmo:OPMGraph .
8       ?graph  opmo:hasArtifact ?a .
9       ?a  rdfs:subClassOf rcto:pCRF .
10      ?a  rcto:hasHCNumber hc .
11      ?a opmo:endTime ?t .
12      ?graph  opmo:hasProcess ?p
13      ?graph  opmo:hasDependence ?dep .
14      ?dep rdfs:subClassOf opmo:WasGeneratedBy .
15      ?dep opmo:effect ?a .
16      ?dep opmo:cause ?p .
17      ?p (opmo:effectInverse/opmo:cause)* ?al .
18      ?al rdfs:subClassOf obi:Person  .
19 }
```

Example 1.4. Queries can also relate to other types of entities. This query returns the document describing the feasibility of the laboratory `lname` and the occurrences of that laboratory in the trial master file.

```
1  PREFIX rdfs: <http://www.w3.org/2000/01/rdf-schema#>
2  PREFIX opmo: <http://openprovenance.org/model/opmo>
3  PREFIX obi: <http://purl.obolibrary.org/obo/obi>
4  PREFIX rcto: <http://www.lesc.imperial.ac.uk/rcto/rcto>
5  SELECT ?doc, ?i
6  WHERE  {
7      ?graph  rdfs:subClassOf opmo:OPMGraph .
8      ?graph  opmo:hasArtifact ?a .
9      ?a  rdfs:subClassOf obi:Laboratory .
10     ?a  obi:hasFacilityName lname .
11     ?graph  opmo:hasArtifact ?doc .
12     ?doc rdfs:subClassOf rcto:FeasibilityAnalysisDocument .
13     ?graph  opmo:hasArtifact ?TMF .
14     ?doc rdfs:subClassOf rcto:TrialMasterFile .
15     ?TMF hasIncidence ?i .
16     ?i rcto:involved ?a .
17 }
```

Example 1.5. Another useful global query gives the listing of all documents used during the execution of a specified process `proc`.

```
1  PREFIX rdfs: <http://www.w3.org/2000/01/rdf-schema#>
2  PREFIX opmo: <http://openprovenance.org/model/opmo>
3  PREFIX rcto: <http://www.lesc.imperial.ac.uk/rcto/rcto>
4  PREFIX rctpo: <http://www.lesc.imperial.ac.uk/rcto/rctpo>
5  SELECT ?doc
6  WHERE  {
7      ?graph  rdfs:subClassOf opmo:OPMGraph .
8      ?graph  opmo:hasProcess ?p .
9      ?a  rdfs:subClassOf PROC .
10     ?p opmo:usedStar ?a .
11     ?a rdfs:subClassOf ?rctpo:DocumentArifact .
12 }
```

Example 1.6. Finally, individual pieces of information may also be retrieved, including which researcher created the linkage data associated to the Serious Adverse Events (SAEs) in the trial.

```
1  PREFIX rdfs: <http://www.w3.org/2000/01/rdf-schema#>
2  PREFIX opmo: <http://openprovenance.org/model/opmo>
3  PREFIX rcto: <http://www.lesc.imperial.ac.uk/rcto/rcto>
4  PREFIX rctpo: <http://www.lesc.imperial.ac.uk/rcto/rctpo>
5  SELECT ?research
6  WHERE  {
7      ?graph rdfs:subClassOf opmo:OPMGraph .
8      ?graph  opmo:hasProcess ?p .
9      ?p  rdfs:subClassOf rctpo:LinkageProcess .
10     ?p rcto:query ?q .
11     ?q rcto:useTable 'SAE' .
12     ?graph  opmo:hasDependence ?dep .
13     ?dep rdfs:subClassOf opmo:WasControlledBy .
14     ?dep opmo:effect ?p .
15     ?dep opmo:cause ?research .
16     ?research opmo:role 'StudyResearch' .
17 }
```

1.6 Summary

Provenance provides a powerful tool for capturing the actions and interactions in complex systems by creating an extensible infrastructure for standardized auditing and in-depth querying of events that led to a certain output or action. Randomized controlled trials in particular are in dire need of both a semantic representation to underpin the knowledge management tools currently in development, and some means of supporting the complex process model proscribed by legal, scientific, and pragmatic requirements. The model presented in this chapter addresses the latter point by delivering a mechanism for standardized auditing of trials using provenance data captured both by software tools and human actors in the process.

The model is the key part of the provenance framework, developed within the context of the TRANSFoRm project to support interpretation, reporting and reproducibility of results delivered by a decision support system built around clinical trial data. The framework performs the capture of all occurrences of data access or queries that occur during studies in the context of the TRANSFORM project, allowing users to track their studies and processes, and reproduce and reconstruct the steps that were used to achieve a particular decision, thereby ensuring trust in the processes and procedures used to perform studies and produce results.

OPM-RCT model was designed as a profile extension to the Open Provenance Model, currently an input to a W3C provenance standard that is in development. The profile incorporates two new ontologies: Randomized Control Trial Provenance Ontology (RCTPO) that specifies additional provenance entities and relations needed to easily formulate important queries within the RCT domain, and Randomized Control Trial Ontology (RCTO) aligning several existing ontologies in related domains, and extending them by RCT specific terms and concepts enabling full representation of ICH Good Clinical Practice guideline on RCTs. The latter ontology is therefore relevant for any RCT knowledge application that wants to conform to the GCP standard.

The general approach taken in the chapter, of defining a semantic representation for the domain, together with a generic provenance module that incorporates the querying capabilities needed, is applicable to any process and document intensive domain, ranging from business enterprises to algorithm flow charts. Indeed, the current direction that the provenance community is taking is the one of developing profile extensions both for fundamental computing concepts (e.g. data models and transformations) and for requirements of particular domains such as document versioning. The eventual aim is to have a rich collection of profiles that can be freely combined to achieve the right level of complexity and expressivity that the domain application requires. OPM-RCT fits firmly within that philosophy.

References

1. Health Level 7. Resource Information Model (2011), `http://www.hl7.org/implement/standards/rim.cfm`
2. European Medicines Agency. ICH Guideline for Good Clinical Practice Topic E6 (R1) (2002), `http://www.ema.europa.eu/pdfs/human/ich/013595en.pdf`

3. Anand, M.K., Bowers, S., McPhillips, T., Ludäscher, B.: Exploring Scientific Workflow Provenance Using Hybrid Queries over Nested Data and Lineage Graphs. In: Winslett, M. (ed.) SSDBM 2009. LNCS, vol. 5566, pp. 237–254. Springer, Heidelberg (2009)
4. Benabdelkader, A.: PLIER: Provenance Layer Infrastructure for e-Science Resources (2011), http://twiki.ipaw.info/bin/view/OPM/PlierToolBox
5. Bishop, B., Kiryakov, A., Ognyanoff, D., Peikov, I., Tashev, Z., Velkov, R.: OWLIM: A family of scalable semantic repositories. Semantic Web Journal 2(1) (2011)
6. Bizer, C.: D2RQ - treating non-RDF databases as virtual RDF graphs. In: 3rd International Semantic Web Conference, ISWC (2004)
7. Broekstra, J., Kampman, A., van Harmelen, F.: Sesame: A Generic Architecture for Storing and Querying RDF and RDF Schema. In: Horrocks, I., Hendler, J. (eds.) ISWC 2002. LNCS, vol. 2342, pp. 54–68. Springer, Heidelberg (2002)
8. Aduna, B.V.: User guide for Sesame 2.3 (2011), http://www.openrdf.org/doc/sesame2/users
9. Chebotko, A., Lu, S., Fei, X., Fotouhi, F.: RDFProv: A relational RDF store for querying and managing scientific workflow provenance. Data & Knowledge Engineering 69(8), 836–865 (2010)
10. OMG Consortium. Unified Modeling Language: Infrastructure, Version 2.0 (2005), http://www.omg.org/spec/UML/2.0/Infrastructure/PDF
11. Medical Research Council. Clinical Trials Toolkit (2004), http://www.ct-toolkit.ac.uk
12. Courtot, M., Bug, W., Gibson, F., Lister, A.L., Malone, J., Schober, D., Brinkman, R., Ruttenberg, A.: The OWL of Biomedical Investigations. In: Dolbear, C., Ruttenberg, A., Sattler, U. (eds.) OWLED, CEUR Workshop Proceedings, vol. 432 (2008), CEUR-WS.org
13. Ellkvist, T., Koop, D., Freire, J., Silva, C.T., Stromback, L.: Using Mediation to Achieve Provenance Interoperability. In: IEEE International Workshop on Scientific Workflows (2009)
14. Orri Erling. Linked Data and Virtuoso (2010), http://www.openlinksw.com/dataspace/oerling/weblog/Orri%20Erling%27s%20Blog/1603
15. Fridsma, D.B., Evans, J., Hastak, S., Mead, C.N.: The BRIDG project: a technical report. Journal of the American Medical Informatics Association: JAMIA 15(2), 130–137 (2008)
16. Harris, S., Lamb, N., Shadbolt, N.: 4store: The Design and Implementation of a Clustered RDF Store. In: 5th International Workshop on Scalable Semantic Web Knowledge Base Systems (SSWS 2009), pp. 81–96 (2009)
17. He, Y., Xiang, Z., Sarntivijai, S., Toldo, L., Ceusters, W.: AEO: a realism-based biomedical ontology for the representation of adverse events. In: Adverse Event Representation Workshop, International Conference on Biomedical Ontologies (ICBO), pp. 26–30 (2011)
18. LP. Hewlett-Packard Development Company. Jena semantic web framework (2009)
19. Idehen, K.U.: OpenLink Virtuoso - Product Value Proposition Overiew (2010), http://www.openlinksw.com/dataspace/kidehen@openlinksw.com/weblog/kidehen@openlinksw.com%27s%20BLOG%20%5B127%5D/1609
20. Franz Inc. AllegroGraph RDFStore (2010), http://agraph.franz.com/allegrograph/
21. Jans, A.: Querying a Trillion triples. In: Semantic Technology Conference (2011)
22. Kifor, T., Varga, L., Vazquez-Salceda, J., Alvarez, S., Willmott, S., Miles, S., Moreau, L.: Provenance in Agent-mediated Healthcare Systems. IEEE Intelligent Systems 21(6), 38–46 (2006)

23. Lim, C., Lu, S., Chebotko, A., Fotouhi, F.: Storing, reasoning, and querying OPM-compliant scientific workflow provenance using relational databases. Future Gener. Comput. Syst. 27, 781–789 (2011)
24. McGuinness, D.L., van Harmelen, F.: OWL Web Ontology Language Overview. W3C Recommendation 10, 2003–2004 (2004) (January 22, 2008)
25. Miles, S.: Mapping attribution metadata to the Open Provenance Model. Future Gener. Comput. Syst. 27, 806–811 (2011)
26. Miles, S., Groth, P., Branco, M., Moreau, L.: The Requirements of Using Provenance in e-Science Experiments. Journal of Grid Computing 5(1), 1–25 (2007)
27. Moreau, L.: The foundations for provenance on the web. Foundations and Trends in Web Science 2(2-3), 99–241 (2010)
28. Moreau, L., Clifford, B., Freire, J., Futrelle, J., Gil, Y., Groth, P., Kwasnikowska, N., Miles, S., Missier, P., Myers, J., Plale, B., Simmhan, Y., Stephan, E., van den Bussche, J.: The Open Provenance Model core specification (v1.1). Future Generation Computer Systems 27(6), 743–756 (2011)
29. Moreau, L., Ding, L., Futrelle, J., Verdejo, D.G., Groth, P., Jewell, M., Miles, S., Missier, P., Pan, J., Zhao, J.: Open Provenance Model (OPM) OWL Specification (October 2010), http://openprovenance.org/model/opmo
30. Murray, C.: Oracle R Database Semantic Technologies Developers Guide 11g Release 2 (11.2). Oracle, E25609-02 (2011)
31. Prud'hommeaux, E., Seaborne, A.: SPARQL Query Language for RDF. Technical report, W3C (2008)
32. Rieb, K., Beauregard, B., Perry, M., Sundara, S., Das, S.: Not Only SPARQL, Not Just RDF/OWL: Special Enterprise-focused Semantic Capabilities for Effective Utilization of Semantic Technologies in Oracle Database. In: Semantic Technology Conference (2011)
33. Salvadores, M., Correndo, G., Harris, S., Gibbins, N., Shadbolt, N.: The Design and Implementation of Minimal RDFS Backward Reasoning in 4store. In: The Semanic Web Research and Applications, pp.139–153 (2011)
34. Shankar, R.D., Martins, S.B., O'Connor, M., Parrish, D.B., Das, A.K.: An ontology-based architecture for integration of clinical trials management applications. In: AMIA Annual Symposium Proceedings, pp. 661–665 (2007)
35. Sim, I., Carini, S., Tu, S., Wynden, R., Pollock, B.H., Mollah, S.A., Gabriel, D., Hagler, H.K., Scheuermann, R.H., Lehmann, H.P., Wittkowski, K.M., Nahm, M., Bakken, S.: The human studies database project: Federating human studies design data using the ontology of clinical research. In: AMIA Summits. Transl. Sci. Proc., pp. 51–55 (2010)
36. Speedie, S.M., Taweel, A., Sim, I., Arvanitis, T.N., Delaney, B., Peterson, K.A.: Model Formulation: The Primary Care Research Object Model (PCROM): A Computable Information Model for Practice-based Primary Care Research. JAMIA 15(5), 661–670 (2008)
37. Tan, W., Madduri, R., Nenadic, A., Soiland-Reyes, S., Sulakhe, D., Foster, I., Goble, C.A.: CaGrid Workflow Toolkit: A taverna based workflow tool for cancer grid. BMC Bionformatics 11 (2010)
38. Thompson, B., Personick, M.: Bigdata: the semantic web on an open source cloud. In: International Semantic Web Conference (2009)
39. W3C. Provenance interchange working group, http://www.w3.org/2011/01/prov-wg-charter
40. Wassink, I., Ooms, M., van der Vet, P.: Designing Workflows on the Fly Using e-BioFlow. In: Baresi, L., Chi, C.-H., Suzuki, J. (eds.) ICSOC-ServiceWave 2009. LNCS, vol. 5900, pp. 470–484. Springer, Heidelberg (2009)
41. Project Website. Mulgara semantic store (2009), http://www.mulgara.org/

42. Wilkinson, K., Sayers, C., Kuno, H., Reynolds, D.: Efficient RDF Storage and Retrieval in Jena 2. In: International Workshop on Semantic Web and Databases (SWDB), pp. 131–150 (2003)
43. Wilkinson, K., Sayers, C., Kuno, H.A., Reynolds, D., Ding, L.: Supporting scalable, persistent Semantic Web applications. IEEE Data Eng. 26(4), 33–39 (2003)
44. Willoughby, C.E., Fridsma, D.B., Chatterjee, L., Speakman, J., Evans, J., Kush, R.D.: A Standard Computable Clinical Trial Protocol: The Role of the BRIDG Model. Drug Information Journal 41(3), 383–392 (2007)
45. Zhao, J.: Open Provenance Model Vocabulary Specification (October 2010), http://open-biomed.sourceforge.net/opmv/ns.html

Chapter 2
Evaluating Workflow Trust Using Hidden Markov Modeling and Provenance Data

Mahsa Naseri and Simone A. Ludwig

Abstract. In service-oriented environments, services with different functionalities are combined in a specific order to provide higher-level functionality. Keeping track of the composition process along with the data transformations and services provides a rich amount of information for later reasoning. This information, which is referred to as provenance, is of great importance and has found its way into areas of computer science such as bioinformatics, database, social, sensor networks, etc. Current exploitation and application of provenance data is limited as provenance systems have been developed mainly for specific applications. Therefore, there is a need for a multi-functional architecture, which is application-independent and can be deployed in any area. In this chapter we describe the multi-functional architecture as well as one component, which we call workflow evaluation. Assessing the trust value of a workflow helps to determine its rate of reliability. Therefore, the trustworthiness of the results of a workflow will be inferred to decide whether the workflow's trust rate should be improved. The improvement can be done by replacing services with low trust levels with services with higher trust levels. We provide a new approach for evaluating workflow trust based on the Hidden Markov Model (HMM). We first present how the workflow trust evaluation can be modeled as a HMM and provide information on how the model and its associated probabilities can be assessed. Then, we investigate the behavior of our model by relaxing the stationary assumption of HMM and present another model based on non-stationary hidden Markov models. We compare the results of the two models and present our conclusions.

Mahsa Naseri
Department of Computer Science, University of Saskatchewan, Canada
e-mail: naseri@cs.usask.ca

Simone A. Ludwig
Department of Computer Science, North Dakota State University, USA
e-mail: simone.ludwig@ndsu.edu

Q. Liu et al. (Eds.): Data Provenance and Data Management in eScience, SCI 426, pp. 35–58.
springerlink.com

2.1 Introduction

In service-oriented environments, services with different functionalities are combined in a specific order to provide higher-level functionality. The composition of services is usually referred to as workflows. A workflow is defined as the automation of the processes and involves the orchestration of a set of services, agents and actors that must be combined together to solve a problem or define a new service. Different services of the workflow represent the transformation processes that receive the data as input to produce the transformed data as output. The workflow graph often describes a network where the nodes are services and the edges represent messages or data streams that channel work or information between services. Each node processes a stream of messages and forwards the resulting streams into its connected nodes.

In such environments, great numbers of workflows are executed to perform mostly scientific and not often business experiments. The workflow activities are run repeatedly by one or more users and large numbers of result data sets in the form of data files and data parameters are produced. As the number of such datasets increases, it becomes difficult to identify and keep track of them. Besides, in these large-scale scientific computations how a result dataset is derived is of great importance as it specifies the amount of reliability that can be placed on the results. Thus, information on data collection, data usage and computational outcome of these workflows provide a rich source of information.

Capturing the execution details of these transformations is a significant advantage for using workflows. The execution details of a workflow, referred to as provenance information, is usually traced automatically and stored in provenance stores. Provenance data contains the data recorded by a workflow engine during a workflow execution. It identifies what data is passed between services, which services are involved, and how results are eventually generated for particular sets of input values. Data associated with a particular service, recorded by the service itself or its provider, is also stored as provenance information. Such data may relate to the accuracy of results a service produces, the number of times a given service has been invoked, or the types of other services that have made use of it [2].

One of the unexplored applications of provenance is exploiting it for the purpose of learning. A large store of the previous executions of services and workflows, as well as their specifications, provide an appropriate data set for learning and knowledge discovery. The provenance data can be explored using data mining and pattern recognition methods to discover the patterns of interest in the data. The store is also a suitable source for learning probabilities. Therefore, probability learning methods can be used to produce the required parameters for the probabilistic decision making processes. As the provenance data is recorded at regular intervals, and consists of values and events that are changing with time, we believe time series mining methods [1] are a suitable choice for evaluating and describing the changes that occur in the data.

Applying learning and knowledge discovery methods to provenance data can provide rich and useful information on workflows and services. Therefore, the challenges with workflows and services are studied to discover the possibilities and benefits of providing solutions by using provenance data. Previously, large amount of research has been done to target workflow challenges such as composition, pattern discovery, service selection, and process refinement. Workflow composition and selection methods require a description of resources and Quality of Service (QoS) specifications as well as well-defined inputs and outputs. These descriptions are usually presented in the service ontologies provided in service registries. As the provenance store keeps the specification of services such as input or output or service description, it can be regarded as a large informational registry providing the chance of intelligent composition and service selection using previous experiences. Among the workflow issues and challenges, workflow analysis and evaluation, which mostly includes QoS assessment and trust measurements, is the least-attended problem. Provenance provides a suitable resource of information for performing analytical evaluation on data. Discovering workflow patterns has been previously studied using event logs, which provide a very small amount of data for learning the workflow models, while provenance provides a rich knowledge base for extracting hidden and unknown models [3].

The remaining sections of this book chapter are organized as follows: in Sections 2 and 3 the motivation and requirements as well as the multi-functional provenance architecture[1] is described. Section 4 outlines how workflow trust can be evaluated using the Hidden Markov Model, in Section 5 we discuss the procedure followed for assessing the HMM probabilities, and in Section 6 the implementation details of the model are provided. Section 7 presents a case study, as well as the stationary assumption of the model is investigated and some experiments are performed to compare the NSHMM trust evaluation results with HMM. In the final section the conclusion and future work is given.

2.2 Motivation and Requirements

A service-oriented architecture provides an environment in which services are shared among distributed systems. Potentially, thousands of services are available, which can be discovered or combined dynamically through appropriate mechanisms for the purpose of workflow selection, composition, or refinement. Thus, current major issues regarding workflow and services can be summarized to service composition and selection, workflow model extraction, refinement, and evaluation. In literature, these problems are targeted via semantic descriptions of services and event logs. In this section, we are going to discuss the knowledge requirements of each problem, and will argue how provenance data satisfies these requirements and provides a suitable platform for improving as well as optimizing the quality of the solutions to these problems. Workflow composition and selection methods require an expressive language that supports flexible descriptions of models and data to facilitate reasoning and automatic discovery and composition. Therefore, they mostly exploit the semantic descriptions of services as well as their QoS specifications from service

[1] These two chapters have been partly published in [2]

repositories or service providers to perform the composition or selection. In [46], the authors discuss the requirements for workflow composition. These requirements can be summarized as follows:

- Workflows must be described at different levels of abstraction that support varying degrees of reuse and adaptation. It is important to mention that this requirement is based on the fact that workflows can often be created by re-using existing workflows with minimal changes.
- Expressive descriptions of workflow components are needed to enable workflow systems to reason about how alternative components are related, the data requirements and products for each component, and any interacting constraints among them.
- Flexible workflow composition approaches are needed that accept partial workflow specifications from users and automatically transform them into executable workflows.

In order to satisfy these requirements, the authors consider three stages for the creation of the workflows, which include: defining workflow templates, creating workflow instances that are execution independent, and creating executable workflows. The three requirements mentioned can be satisfied through provenance data. In [47], the authors argue that a robust provenance trace provides multiple layered presentation of provenance. Thus, a layered architecture and engine for automatically generating and managing workflow provenance data is considered in provenance systems. As a result, provenance data can be used for interpreting the services and datasets of the workflows. Provenance creation is performed by following a layered approach that fulfills the requirements of the workflow composition process. The first layer of the architecture represents an abstract description of the workflow that consists of abstract activities with the relationships that exist among them. The second layer provides an instance of the abstract model by presenting bindings and instances of the activities. The third layer captures provenance of the execution of the workflow including specification of services and run-time parameters. The final level captures execution time specific parameters including information about internal state of the activities, machines used for running, status and execution time of the activities.

As the execution time specific parameters are also gathered in provenance stores, provenance data also includes the QoS specifications of services. Thus, service selection solutions can be applied to this data in order to automatically select appropriate services that provide some QoS requirements. Service providers may not be trustworthy enough to deliver the services based on the agreed-on QoS. On the other hand, the *validity period* of the agreement might have come to an end and no agreement updates might have been made afterwards. The ontological QoS specification of service providers are updated periodically while there might be many requests in each period. In case the QoS guarantees change during a period, the providers will not be able to satisfy the agreed-on thresholds. Or the service provider might not be able to provide the specifications at all. Using the history of previous executions, the provided QoS overcomes the inconsistencies between the guaranteed and delivered QoS values of services to some extent by providing an estimate of the QoS parameters of the services with regard to time.

Most research on workflow systems focus on prediction, tracking and monitoring of workflows, and not on the evaluation of these processes. Few research efforts which studied the evaluation component of workflows, investigated a very narrow research problem aimed to improve the performance or fault tolerance of workflow systems [6]. As the provenance information maintains the records of previous execution details of workflows, it provides the facility to analyze, assess, and evaluate the behavior of a workflow as well as its performance. The performance of a workflow, its trustworthiness, improvements, and its future trend, etc. can be analyzed and evaluated through provenance data.

Workflow mining discusses techniques for acquiring a workflow model from a workflow log. Workflows can be investigated from many perspectives: functional, behavioral, informational, organizational and operational. In case of the behavioral perspective, which looks at control flow, workflow mining is done by following the order in which events for tasks are stored; for the informational perspective which looks for data flow, usually inputs/outputs are being used; in case of the organizational perspective, participants of tasks and their roles are being discovered. The workflow mining methods use the event-logs for discovering the patterns and mining the workflows, which keep track of a very small amount of information. The information provided in event logs is not enough for mining workflows with regard to all the mentioned workflow perspectives while much stronger reasoning and mining can be done over the data presented in workflow provenance.

To improve the efficiency of the composition and selection processes, previous executions of workflows and services can be used to augment these processes with more intelligence during the composition or selection. The feedback learned through previous runs secure the composition (or selection) from services that either do not have available resources, or do not satisfy the promised trust levels at a particular time. In case of the composition, the feedback of previous runs of the composed process will also be analyzed later to discover the possible deficiencies that might exist in the composed model.

As more provenance information is gathered, the extracted workflow process models are refined over time and the structure is geared to improve the efficiency with regard to changes in the data. These variations might include updates of the most frequently chosen paths, or assigning/changing the weights of the links in the model with regard to the rate of usage in time. These types of augmentations in the model also facilitate the process of refining or repairing a workflow model.

Since the provenance information of the same executions might provide the intermediate data generated by a process, the processes can be reduced by removing existing services, or replacing the parts, which cannot be executed with other parts, by looking for a more optimal path in the extracted workflow model with regards to the weights of the connections.

As mentioned earlier, the history of previous executions of workflows and services satisfies the requirements of addressing the discussed challenges. Apart from the requirements, it was discussed that the provenance data augments the challenges with more efficiency, and reliability. Thus, there is a need for an architecture that facilitates addressing and solving all these aforementioned issues by exploiting the provenance data.

2.3 Architecture

In this section, the multi-functional architecture discussed earlier is presented along with its components. Figure 2.1 outlines the architecture. The structure is composed of 5 components that cooperate together along with the provenance store to provide different functionalities. The responsibilities of each component, the way components collaborate to provide the promised functionalities, and the approach taken to achieve the goals of the components are discussed.

Workflow Model Extraction and Discovery Component: This component is responsible for extracting the workflow pattern and associations that exist among the relevant workflows previously run and executed. Two workflows are considered relevant if they are in the same area of interest. The extraction component discovers the hidden connections that might exist among services and were not known beforehand. It generates a policy graph of the relevant services with edges representing the associations between them. The output is an optimal policy graph including all possible paths that could exist between the services of similar functionality. The extracted policy graph can be used later for the purpose of workflow construction and repair. The component is also able to receive a workflow pattern, and look for the same pattern sequence in the store to discover if there is any information regarding its previous executions in the provenance store.

Workflow and Service Evaluation Component: Evaluating workflows and services in terms of trust and quality is an important and less studied topic in the area of workflows. Workflows need to be assessed and analyzed to discover how trustworthy the composition of services are, therefore, in case the trust given by a workflow is not satisfactory, the workflow sequence can be repaired and improved. Another responsibility of this component is to identify the points in time at which a significant variation in trust occurs. This information can help us in identifying the parts of the workflow that are not providing the promised or required trust levels. Similar to workflows, the services are evaluated by this component. Large fluctuations of the QoS values of services are investigated to predict when in the future the service will not support the promised QoS requirements. Based on the previous executions, this component is also able to predict which services are going to be executed and in case the results of another instance of the same service are available, the process of workflow execution can be improved by exploiting those results. Apart from the trust assessment, the performance of the workflow is evaluated in terms of resource usage, and total time elapsed from the submission to completion.

Workflow Repair and Refinement Component: In case a workflow does not provide the required trust level, or it cannot be executed due to lack of available services, the workflow needs to be repaired or refined. The repairment/refinement component takes advantage of the extracted policy graph of the workflow along with the assessment results of the evaluation component. The policy graph is traced to find a path that can replace the defective part of the workflow. The defective path is either inefficient due to lack of trust provision, or cannot be executed any longer because of unavailable services. In case a service is predicted to not provide the promised non-functional requirements, the service is replaced by another service or services to provide a similar functionality.

Workflow Composition and Generation Component: Composing a set of services using provenance data is a very useful exploitation of the provenance store. The stored specifications of services and their states provide the facility of composing the services automatically. On the other hand, having the previous history of executions, provides the data, which is essential for learning, therefore, the composition will be done in a more efficient way by exploiting the provenance data. This component receives the requirements and composes a workflow dynamically by taking advantage of the service specifications provided in the store. Previous execution of workflows enables the composition to be more robust as it exploits the evaluation results of services and workflows to generate a well-designed workflow process.

Workflow Service Selection Component: The problem of selecting a set of concrete services that provide the required QoS specifications for a complete abstract workflow is referred to as abstract workflow service selection problem. The provenance data can be exploited to speed up this task. In order to find the set of concrete services that match a single abstract service, service registries are looked at and matchmaking algorithms are applied to discover the matching services. The service

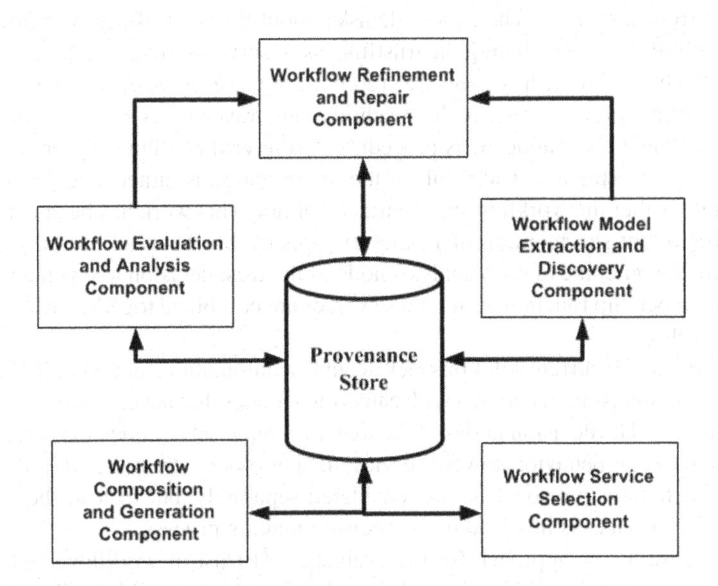

Fig. 2.1 Architecture

discovery phase is much simpler if provenance data is used. Previous executions of workflows along with the workflow templates simplify the process of service discovery for a simple query. The set of suitable concrete services for the abstract workflow can then be selected more optimally by using the selection mechanisms along with the evaluations of previous executions.

2.4 Hidden Markov Modeling for the Evaluation of Workflow Trust

In the remainder of this book chapter we want to focus on the workflow evaluation component of the architecture.

Execution of a sequence of services requires much more resources and time in comparison to a single service. Thus, if a workflow is not very reliable, many resources and time will be wasted; since the results of the workflow can not be trusted. Therefore, it is important to be able to evaluate the trust of a workflow to find the degree of reliability of the workflow and its results. This also helps to decide whether the workflow needs some refinement and whether less trustful services should be exchanged with more trustful ones.

Having the trust value of each service, allows to evaluate the overall trust value of a sequence of services, i.e. a workflow. Therefore, we can determine the amount of trust that can be placed on the overall workflow as well as the results and datasets generated during the workflow execution. There are very few approaches addressing the subject of workflow trust evaluation. One approach uses a decision tree model, which is presented in [19]. In this chapter, a decision tree is built out of a question sequence that will help in assessing the trust that can be associated with the data produced from a process. The root node asks about the trust of the workflow and has three child nodes, evaluating the trustfulness of services, data and the workflow process. Each child node has a sub-tree representing a set of yes/no questions. The decision making process starts with one child node, traverses its sub-trees and continues to the next child node. This procedure is followed continuously until all the sub-trees are investigated. The result of the investigation is either a yes or no, determining whether the workflow can be trusted or not. This work has been extended and an important shortcoming of it, the crisp result, has been addressed in [20]. Therefore, the outcome of each analysis node of the trust decision tree is mapped to a fuzzy membership function. Later, these values are combined together using fuzzy inference rules.

However, all the current solutions lack accuracy, automation, and reliability. They are based on a decision tree model with categorical nodes that have been designed by the developers. The decision nodes of the tree are simple sets of questions regarding the user's views or behaviors toward service, data or process trust. Besides, the trust value of each service or data is not considered separately, but instead the overall trust level of services is involved in the decision making process.

We propose a new approach for the evaluation of trust of workflows, which is based on a statistical model named Hidden Markov Model (HMM). Rather than traversing a set of question nodes, in our model, the trust will be assessed by solving a set of mathematical equations that describe the behavior of the workflow trust in terms of random variables and their probability distributions. Thus, our method is more accurate in comparison to the previous approaches and will support automation.

A HMM is a probabilistic process over a finite set of states, where each state generates an observation. Given a HMM, and a sequence of observations, the probability of the observation sequence given the model can be evaluated. It is also possible to discover the hidden state sequence that was most likely to have produced the observation sequence. Another type of inference on HMMs can estimate the HMM model through training examples and learning methods.

HMM has become the method of choice for modeling stochastic processes and sequences in applications such as speech and handwriting recognition [8], computational molecular biology [9], natural language modeling [10], etc. In this work, HMM is used for the purpose of workflow trust evaluation.

In order to be able to assess the proposed HMM model, probability learning algorithms like Maximum Likelihood (ML) or Expectation Maximization (EM) learning techniques are used along with provenance data. Provenance is one of the growing demands in distributed service oriented environments, which supports the systems with documentation of the origin and the processing steps of data that is part of a workflow execution process. It also provides explanations about which, how and what resources and services were used to produce that data, and is referred to as provenance data that is captured and stored in provenance stores for the purposes of reasoning, validation and re-execution. A provenance store provides the necessary information that is exploited for the purpose of estimating HMM probabilities.

Many approaches have been proposed to improve the predictive power of HMM in practice. For example, factorial HMM [12] is proposed to decompose the hidden state representation into multiple independent Markov chains. In speech recognition, factorial HMM can help in representing the combination of multiple signals. Hierarchical HMM [13] is another method that facilitates the inference of correlated observations over long periods in the observation sequence via higher level hierarchy. However, from the essential definition of HMM, there are other ways to improve the predictive power of HMMs. One approach is to relax the stationary hypothesis of HMMs and make use of time information. To investigate this further and observe the behavior of our model with regard to the non-stationary assumption, the workflow trust has also been evaluated using the Non-Stationary HMMs (NSHMM).

2.5 Methodology

The notion of trust of an enacted workflow is an important issue in distributed service oriented environments. Trust evaluation aims at contributing in the discovery of how trustful the results of a workflow are. It also helps the optimization of composite service executions. In this section, we are going to first present how the workflow trust can be evaluated using hidden Markov modeling. Later, we explain how the model can be assessed by taking advantage of the previous history of the execution of workflows.

A HMM is a statistical model that can be considered as the simplest kind of Dynamic Bayesian Networks (DBNs). The system that is being modeled according to HMMs is assumed to be a Markov process with unknown parameters. Markov processes are an important class of stochastic processes that are governed by the Markov property. The Markov property states that the future behavior of a process given its path only depends on its present state. The HMM model basically consists of two sets of variables: state variables and evidence variables, which are also called the observations. The state variables are the hidden variables that change over time; while the evidence variables are the observable variables that are known in advance at each time step. The challenge is to determine the hidden parameters from the observed ones.

Figure 2.2 shows a simple first order HMM. The state variable x_t is a hidden variable at time t and can have a value from x_t the domain of x. The random variable y_t denotes the observable parameter at time t. From the figure, it can be seen that the value of the hidden variable at time t, i.e. x_t, depends only on the value of the hidden variable x_{t-1}, and other previous parameters have no influence on it. This property is referred to as the first order Markov property.

In order to model the workflow trust evaluation as a HMM, the state and observable variables are mapped as follows:

- Tr_t: the trust state variable, represents the state of the trust of the workflow at time t.
- S_t: the evidence variable represents the service that is being executed at time t.

Figure 2.3 depicts a simple linear workflow and the correspondent HMM, modeled to evaluate the trust level of the workflow. As it can be observed form the figure, the state of the trust of the workflow at the beginning (Tr_0) is only determined by the evidence variable observed at that time (x_t). For the following time steps, the state of the trust of the workflow can be determined by investigating the state of the workflow at the previous time step, and observing the service that was executed at that time.

In theory of HMMs, some assumptions are made for the sake of mathematical and computational tractability. Here we present how these assumptions can be applied to our model:

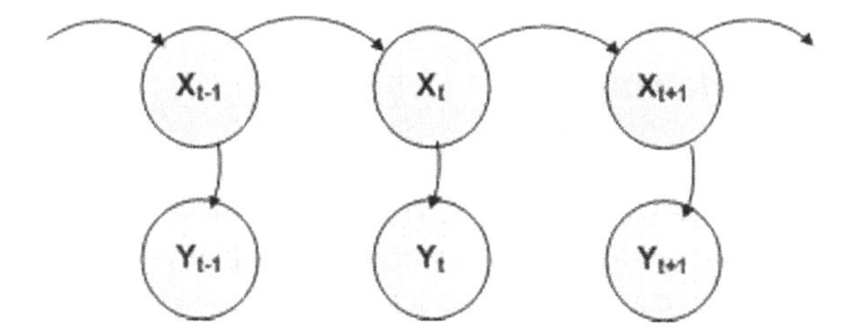

Fig. 2.2 Basic HMM.

1. The Markov assumption: It is assumed that the next state is dependent only upon the current state. This is true in case of our model, as the state of the trust of the workflow at each time only depends on the state of the trust at the previous time and not the other prior states.
2. The output independence assumption: This is the assumption that the current observation is statistically independent of the previous observations. In case of our model, the service at time t is independent of the previous services.
3. The stationary assumption: This assumption is based on the fact that the transition probabilities between the states are independent of the actual time at which the transitions take place. In case of the workflow trust problem, we can not say that transition probabilities are completely independent of time. We suppose that this assumption will be true for our model since we can take the average of the state transitions of all times and have one set of state transition probabilities for the overall time period. In order to investigate this further, later in the chapter, we will observe the behavior of the model by relaxing this assumption and having a non-stationary HMM.

Having defined the HMM and described how the HMM parameters and assumptions can be mapped to the workflow trust evaluation parameters, we will now clarify how this model can be exploited for the purpose of trust evaluation.

As mentioned earlier, different kinds of inference can be done on HMM structures. These include methods for computing the posterior distribution over the current, future, or a past state, or finding the sequence of states that is most likely to have generated those observations. Filtering or monitoring is the task of computing the posterior distribution over the current state, given all evidences and observations to date. The following probability expresses filtering inference:

$$P(X_t \mid y_1, y_2, \ldots, y_t) \tag{2.1}$$

Using the filtering model, the probability of the state of the trust at the final state of the workflow can be roughly estimated given all the observations, which are the services seen so far. Therefore, for the case of the trust evaluation, the following probability should be assessed:

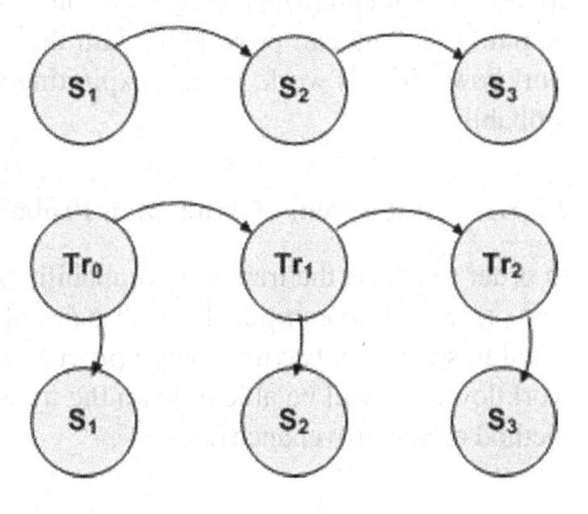

Fig. 2.3 A sample workflow and the HMM for workflow trust evaluation.

$$P(Tr_2 \mid s_1, s_2, s_3) \tag{2.2}$$

for different possible trust state levels. Evaluation of the above probability provides us with estimations of probabilities for different trust levels at time t_2. In this work, the state of the trust will be evaluated at three different levels of *High*, *Medium* and *Low*. The work can later be extended to support further trust levels.

2.5.1 *Trust Model Assessment*

In order to be able to compute the filtering inference, two other probabilities should be assessed beforehand. These probabilities are referred to as state transition probability and sensor probability. The state transition probability is defined as the probability of being in the next state given the current state, i.e. $P(x_t \mid x_{t-1})$, which in our case is the probability of being at a trust level at time t given the level at the previous time, i.e. $t-1$. The sensor probability is defined as the probability of the observation at time t, which is the service that was executed at time t, given the different level of trustworthiness of the workflow at that time. To assess the state transition or sensor probabilities, the ML or EM learning algorithms are utilized along with the provenance data.

In service-oriented environments, great numbers of workflows are executed to perform computational and business experiments. The workflow activities are run repeatedly by one or more users and large numbers of result data sets in the form of data files and data parameters are produced. As the number of such datasets increases, it becomes difficult to identify and keep track of them. Besides, in these large scale scientific computations how a result dataset is derived is of great importance as it can specify the amount of reliability that can be placed on the results. Thus, information on data collection, data usage and computational outcome of these workflows provide a rich source of information. Capturing this information, which is regarded as provenance information, is a significant advantage of using workflows. Provenance information facilitates data dependency determination, workflow result validation, efficient workflow re-executions, error recovery, etc. [14]. Provenance also enables users to trace how a particular result has been arrived at by identifying the aggregation of services that produces such a particular output. This data can provide us with the history of previous execution details of workflows. In this work, we are exploiting the provenance data to learn the HMM probabilities.

2.5.1.1 Assessment of Transition Probabilities

In order to assess the transition probabilities, the trust state transitions, i.e. $P(Tr_t \mid Tr_{t-1})$, should be computed for all pairs of workflow services that are being executed in sequence. Having a large provenance record of the previous executions of workflows, we will be able to learn the transition probabilities by applying the ML method on the provenance data.

ML learning is a data analysis approach for determining the parameters that maximize the probability (likelihood) of the sample data, which is trust state transitions in this case. From a statistical point of view, the method of ML is considered to be robust and yields probabilities with good statistical properties [15].

To assess this probability using the ML method, we determine the number of each trust state transition with regard to the total number of transitions of that state. The transition probability estimation for our model is computed based on Equation 2.3:

$$P(Tr_t = j \mid Tr_{t-1} = i) = \frac{n_{ij}}{n_i} \tag{2.3}$$

where n_{ij} denotes the number of transitions from trust level i to trust level j, and n_i denotes the number of transitions from trust level i. For example, for the sample workflow in Figure 2.3, which was composed of three services, the trust state transition from high to low will be computed by first determining the number of high to low transitions for the service pairs (s_1, s_2) and dividing it by the number of times the service s_2 had low trust level. The same will be done for the pair (s_2, s_3). The average of these values represents the transition probability from high to low.

It is important to mention that the same pair of sequential services might be repeated in several workflows, and the transition probabilities for these services will be learnt without considering specific workflows. The average of all these probabilities will denote the final transition probability for these pairs of services.

Assessment of Sensor Probabilities

To assess the sensor probabilities for each time instance t, the probability of observing an evidence variable given the state at that time should be computed. Therefore, we should compute $P(S_t \mid Tr_t)$, which again will be learnt by utilizing the ML method and the provenance data.

For this purpose, the number of times the trust state of service instance S_t was at each trust level is estimated. This value is divided by the total number of times any service was at that trust state. As before, the provenance history of the workflow will be used. Equation 2.4 represents the assessment of the sensor probabilities for our model:

$$P(S_t = s_t \mid Tr_t = j) = \frac{n_{stj}}{n_j} \tag{2.4}$$

where n_{stj} denotes the number of times being in state j and observing service s_t, and n_j denotes the number of times being in state j.

Assessing the Trust Level

Having assessed the sensor and transition probabilities, we will be able to assess the filtering model of HMM and therefore evaluate the workflow trust using Equation 2.5:

$$P(Tr_t \mid S_1 = s_1, S_2 = s_2, ..., S_t = s_t) = \propto P(S_t = s_t \mid Tr_t) \sum_t (P_{Tr} \mid P(Tr_{t-1})$$
$$P(Tr_{t-1} \mid S_1 = s_1, S_2 = s_2, ..., S_{t-1} = s_{t-1}) \tag{2.5}$$

The probability of $P(Tr_{t-1} \mid S_1 = s_1, S_2 = s_2, ..., S_{t-1} = s_{t-1})$ is computed recursively. Equation 2.6 evaluates the probability of different trust levels at time t having observed the services the workflow is composed of until that time.

As discussed, for the purpose of assessing the probabilities, the ML learning algorithm is utilized in this work. This is based on the assumption that the provenance data does not include a large amount of missing data. To be able to find the probabilities in case of missing data, the EM learning algorithm can be used. The EM algorithm is an efficient iterative procedure to compute the ML estimate in the presence of missing or hidden data. Using this algorithm, we first predict the missing values based on assumed values for the parameters. Later, these predictions are used to update the parameter estimates. The sequence of parameters converges to ML estimates, and EM implicitly averages over the distribution of the missing values.

2.5.2 Cases with Dynamic or Parallel Sections

The presented trust model is compatible for workflows which contain not only sequential but also parallel sections in the workflow. In case of non-sequential workflows, a sequential workflow is extracted from them by selecting one of the subsections of each parallel section according to a policy, and replacing that parallel subsection with the selected subsection. Starting with the deepest parallel subsections, a subsection is chosen for each section by first applying the HMM model to all the parallel sub-sections of that section, and then the trust level probabilities of the subsections are compared with each other. For each section, the subsection that has the lowest trust level is selected and the parallel section is replaced by that subsection. By following this policy for all the parallel sections, the workflow is transformed to a sequential workflow, and finally the HMM model is applied to assess the trust level.

It is important to mention that as the proposed approach exploits provenance information to get an assessment of the QoS values, it works for the static scenarios. In case of workflows with services for which there is no history in the provenance store, the online QoS values presented by the service provider are used for assessment.

2.6 Implementation

As mentioned earlier in this work, the trust of each service instance is categorized into three levels of *High*, *Medium*, and *Low* and can be evaluated by aggregating the QoS parameters of the service. These QoS parameters can include status, availability, reliability, execution time, reputation, etc. The trust value is usually determined by assigning a weight to each parameter and the summation of the multiplication of the parameters by their weights results in the final trust value. As in our current model we are concerned with trust levels rather than trust values, we determine the level of the trust with regard to the level of the QoS parameters.

In our implementation, we have considered the QoS parameters of status, reliability and availability. The QoS parameter *status* is a binary value that represents the status of the execution of the service. A value of 1 describes that the service was executed successfully and a value of 0 reports unsuccessful execution. The QoS parameter availability presents how available a certain service and its data are, while reliability denotes the degree we can rely on the processing and the response time of the service. Both parameters have a value in the range of [0,1].

In order to decide about the trust level of each service using these parameters, we followed a table model, Table 2.1, in which the level of all QoS parameters of availability and reliability in conjunction with the status of the execution determines the level of the trust. The table is referred to as the trust level decision table throughout this book chapter. A sample row in this table represents the associated trust level in combination with the discussed QoS parameters. For example, LL1 denotes that the level of the reliability and availability of a service is *Low*, and the status is 1. According to the table, the trust level of the service is assessed as *Low*.

The levels of *reliability* and *availability* of the services are determined according to a set of pre-determined range levels. For the examples and experiments provided in this book chapter, the following range table (Table 2.2) was used.

As was discussed earlier, the probabilities are assessed by applying learning methods over the provenance data. For the purpose of learning, we implemented a provenance store in MySQL [28] including tables for storing the information of workflows, services, workflow instances, and workflow sequences. The provenance data is then generated by a random workflow generator implemented to produce instances of a workflow. The generator asks for the following parameters as input:

- N_s: the number of services the workflow should be composed of.
- N_w: the number of previously executed instances of the workflow.

In order to assess the HMM, we followed the matrix algorithm which describes the sensor and transition models in form of matrices. The transition matrix denoted by T is a $m \times m$ (in our case 3×3) matrix where m is the number of possible states. The probability of a transition from state i to state j is denoted by the entry T_{ij}:

$$T_{ij} = P(Tr_t = j \mid Tr_{t-1} = i) \tag{2.6}$$

which, as discussed, will be evaluated using the generated provenance data along with the trust level decision table (Table 2.1), QoS parameters range level (Table 2.2) and the ML algorithm.

The sensor model is also put into matrix form. For each time step t, a diagonal matrix, O_t, is constructed whose diagonal entries are given by the values $P(S_t \mid Tr_t = i)$, with the other entities set to 0.

Now, to accomplish the filtering inference and represent the forward messaging in HMMs using the matrix model, Equation 2.7 is applied recursively:

$$f_{1:t+1} = \alpha O_{t+1} T^T f_{1:t} \tag{2.7}$$

Table 2.1 Trust level decision table, L, M, and H denote Low, Medium, and High.

Trust	Reliability, Availability, Status
L	LL0
L	LL1
L	ML0
M	ML1
L	HL0
M	HL1
L	LM0
M	LM1
L	MM0
M	MM1
L	HM0
H	HM1
L	LH0
M	LH1
M	MH0
H	MH1
M	HH0
H	HH1

Table 2.2 Range Level of the QoS parameters Availability, and Reliability.

Trust Level	Low	Medium	High
Availability	[0,0.3]	(0.3,0.7)	[0.7,1]
Reliability	[0,0.3]	(0.3,0.7)	[0.7,1]

where α is the normalization factor. The result is a one column matrix denoting the probability of the trust level of the workflow for all the different possible levels.

2.6.1 Verification of the Model

Our approach is verified by a comparison done with the Viterbi algorithm [17], which finds the most likely sequence of hidden states that result in a sequence of observed events. For the verification, different observation sequences of different sizes were generated and the most likely sequence of underlying hidden states that might have generated those observation sequences was produced by applying the Viterbi algorithm. Having compared the resulting hidden states of the algorithm with the real hidden states, we received identical results. Therefore, this verifies that the HMM modeled for the purpose of workflow trust evaluation and the way the probabilities were assessed is valid.

2.7 Case Study

In this section, we present a workflow scenario and describe how its trust can be evaluated using the presented model. The sample workflow is the process of knowledge discovery in databases which is referred to as KDD process [16]. The KDD process is composed of four services for data selection and cleaning, data transformation, data mining, and data interpretation. Figure 2.4 shows the process.

The following assumption is made. A distributed service-oriented environment is sharing services for the purpose of knowledge discovery, and that a workflow is executed using four different services shared by service providers in the environment each having different QoS values, and therefore, different trust estimations. Using the workflow generator, the above workflow was defined and 50 execution instances were generated, representing the provenance data. Table 2.3 shows the average of the QoS parameters of those instances.

The QoS parameters *availability* and *reliability* were generated in the range of 0.3 to 0.9, which mostly covers the medium and high trust levels. The status of the execution was set to zero in less than 20% of the cases. It is important to emphasize that according to the trust level decision table (Table 2.1) the state of the trust of a service instance is evaluated as *Low* if its status is zero. The reason for this decision is that if a service does not complete its execution successfully, that service instance should not be trusted at all. Therefore, we evaluate the trust as low regardless of the instance's level of *reliability* and *availability*.

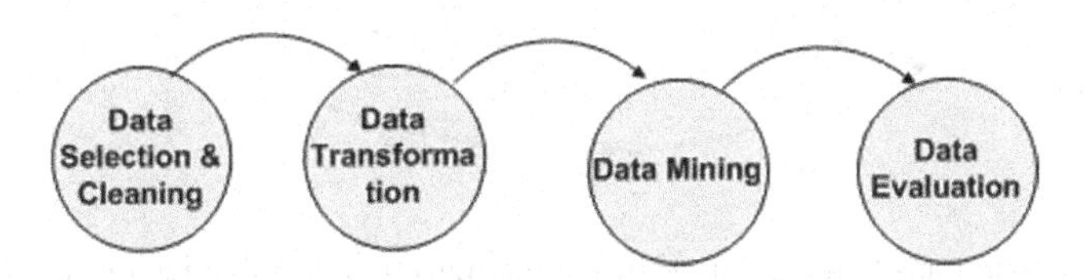

Fig. 2.4 A sample workflow scenario - KDD Process.

Table 2.3 The average of the values of the QoS parameters generated for the scenario.

QoS Parameter	Reliability	Availability	Status
Data Selection	0.58	0.59	0.8
Data Transformation	0.7	0.7	0.88
Data mining	0.34	0.34	0.82
Interpretation	0.84	0.84	0.82

In the next step, the transition matrix is built by learning the probabilities from the generated provenance data. Given the data, the transition matrix T, of the above example was estimated as given in Figure 2.5.

L, M, and H represent the trust levels *Low*, *Medium* and *High*. An entry T_{ij} denotes the transition probability of being transferred from trust level i to j. For a better understanding, the state transition diagram is also provided in Figure 2.6, which is the same as the transition matrix but presents it in a graphical view which is easier to follow.

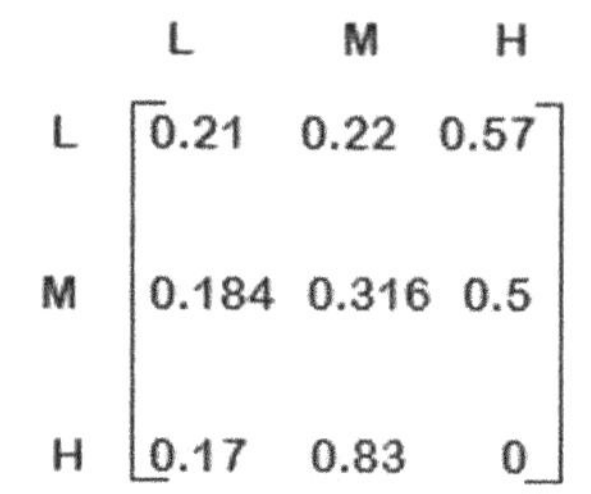

Fig. 2.5 Transition matrix of the example

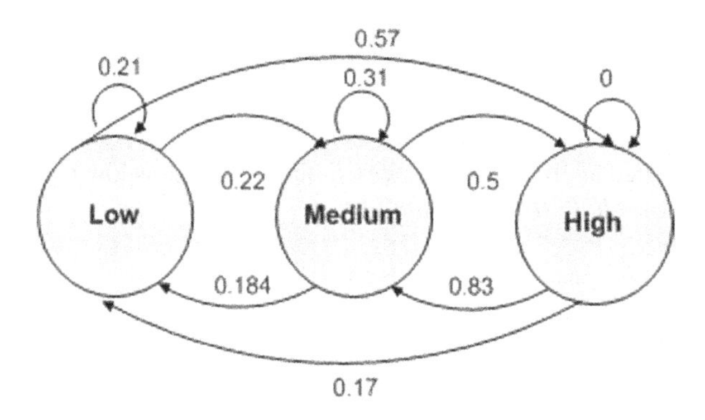

Fig. 2.6 The state transition diagram showing the transition probabilities for the above example learnt through ML method.

Having learnt the transition matrix, the forward algorithm starts with assessing the sensor probability at the first time step and forwards this message along with the transition messages to the next time step. This process of forwarding messages continues until the last service is observed, and therefore the overall trust of the workflow is evaluated. It is important to mention that the prior belief about the trust state probabilities, i.e. the initial state probabilities, is considered equal for all the three possible states and was set to 0.33 for all the trust levels.

To investigate the behavior of the filtering method and observe the trust level probabilities estimated at each time step is provided in Figure 2.6. The figure shows how the trust state probabilities change over time during the HMM assessment for the discussed example.

It can be observed that the trust state is evaluated as *Medium* after observing the first service, it then heads toward *High*, then again *Medium* and finally the trust level is evaluated as *High*.

Taking a look at the average values of the QoS parameters of each service explains the behavior of the model. According to the QoS range evaluation table (Table 2.3), the trust level of the first service, which is the data selection service, can be evaluated as *Medium*. The trust level of the third service is also evaluated as *Medium*, and the trust level of the second and the fourth service is estimated as *High*.

The explanation above and the transition matrix shown in Figure 2.5 describe the reason behind the path taken in Figure 2.7. The path shows the route between the trust levels with the highest probabilities at each time step. The transition probabilities with large probability values include transitions from *High* to *Medium*, *Low* to *High*, and *Medium* to *High*. The evaluation process starts with the first service which has an average of *Medium* trust level. As the transition probability of *Medium* to *High* is the largest, this leads the state of the trust toward *High*. Being in state *High* and having observed a service with *High* trust level leads the trust level toward *Medium* as the largest transition probability from *High* is the one toward *Medium*. The rest of the transitions can be explained in the same way.

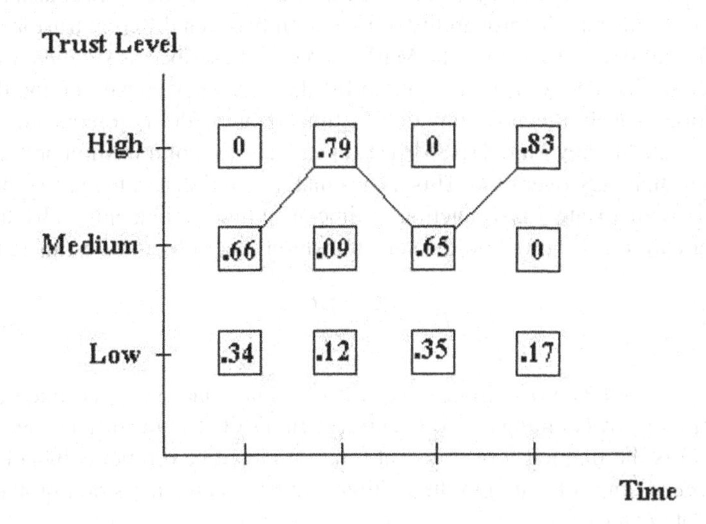

Fig. 2.7 The change of the trust state probabilities over time using a HMM.

It should be considered that there is always a less than 20% probability for a low trust state to be chosen for all the services. Because as discussed earlier, the status of the executions of services were randomly set to zero in almost 10 to 20 percent of the cases. Therefore, the final trust level probabilities will have a 10% low level probability on average.

2.7.1 Investigation of the Stationary Assumption

It was mentioned earlier that one of the assumptions of HMM is the stationary assumption. In order to follow this assumption, the transition probabilities were assessed by taking the average of the transitions between each pair of services to have the same state transition matrix at all times. As this assumption can not be verified completely in case of the workflow trust problem, this section investigates how the model will behave if we relax this assumption and transition probabilities are considered time-dependent. To achieve this goal, the transition probabilities are computed separately for each time step.

In the theory of HMMs, it is assumed that state transition probabilities are independent of the actual time at which the transitions take place. This assumption can be mathematically presented as:

$$P(x_{t1+1} = j \mid x_{t1} = i) = P(x_{t2+1} = j \mid x_{t2} = i) \tag{2.8}$$

for any t_1 and t_2. Equation 2.8 states that the transition probabilities are constant over time which means that the probability of transition between different trust levels is the same for all times. Therefore, the Markov chain is described as *stationary* in the strictest sense. In general, it is possible to lift the constancy constraint and define the transition probabilities as a function of time. This model is referred to as the Non-Stationary Markov Model (NSMM) [11] and has a set of transition probability distributions that vary over time. This means that, given a state i, the probability of moving to another state j is a function of time. The time can be either absolute or relative. Equation 2.9 shows how the state transition function can be estimated:

$$P_{ijt} = \frac{C(i,j,t)}{C(i,t)} \tag{2.9}$$

where $C(i,j,t)$ is the co-occurrence frequency of state i and state j at time t and it can be estimated by counting the co-occurrence times of state i and state j at the tth time. $C(i,t)$ is the frequency of state i at time t and can be estimated by counting the occurrence times of state i in the tth time. And P_{ijt} is the transition probability between state i and j at time t.

In case of the workflow trust evaluation, the transition probabilities can be considered as a function of time since the probability of transition from one trust level to the other at time t depends on the services that are being executed at that time instance. Therefore, it is important to investigate the behavior of the model this time using the NSHMM in order to observe the effect of the stationary assumption on the trust evaluation results.

In case of relaxing the stationary assumption for the workflow trust evaluation, the state transition probabilities were assessed separately at each time step and a transition matrix was built using the ML method along with the provenance data representing the history of the observations seen previously at those time steps.

Following the ML estimation method, the transition probability from state i to state j at time t will be assessed as follows:

$$Pt(Tr_t = j \mid Tr_{t-1} = i) = \frac{n_{ijt}}{n_{it}} \tag{2.10}$$

where n_{ijt} denotes the number of transitions from trust level i to trust level j at time t, and n_{it} denotes the number of transitions from trust level i at time t.

The non-stationary model was further implemented and the result of the same scenario studied in the previous section was investigated using the new model. It is observed that the trust state probabilities have not changed much as time elapses. The maximum trust level path follows the same routine with very little changes in the state probabilities at each time. The evaluation result of the NSHMM shows that the workflow can be trusted with a probability of 93%, while using the HMM this probability was 83%.

To investigate this further, we ran experiments using both models and compared their results. The experiments were done by creating workflows with 5 to 25 services in increments of 5. A previous execution history of 50 instances was randomly generated for each workflow in order to learn the sensor and transition probabilities. The average of the trust level probabilities was then computed. It was observed from the experiment results that for both models the distance between the same trust levels was equal in 96% of the cases.

Figure 2.8 represents the average trust level probabilities of the HMM compared to NSHMM. It can be observed that the differences are very small. In all the experiments, the level of the trust was estimated to be the same.

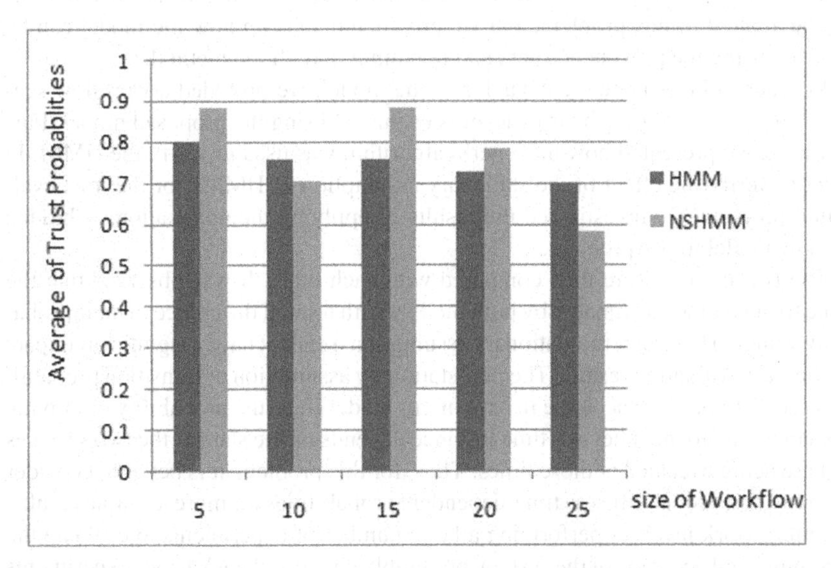

Fig. 2.8 Comparing the average trust level of HMM vs. NSHMM for 5 to 25 numbers of services with increments of 5.

In order to determine whether the results of the two models are the same, we ran the paired T-test on the datasets of the two models. The T-test is a statistical test that assesses whether the means of two groups of data are statistically different from each other. The result was a p-value of 0.78, which represents that the datasets are not significantly different from each other. The chart in Figure 2.8 and the T-test results both verify that the stationary assumption does not have a significant effect on the results of the trust level assessment, as both models provide estimations for the same trust levels with very little difference.

Experiments were done to compare both models in terms of the execution time and it was observed that while there is not large differences between the execution times, the execution time of the non-stationary model is larger. The reason for this observation goes back to the transition matrices that should be computed for each time instance separately while for the HMM with stationary assumption, the transition matrix is built once at the beginning by computing the average of all values.

2.8 Conclusion and Future Work

In this book chapter, a multi-functional architecture was described that addresses the current research issues of workflows and services using provenance data. The components of the architecture were described consisting of model extraction and discovery, workflow evaluation, workflow repair and refinement, workflow composition, and workflow service selection.

In addition, we focused on one component of the multi-functional architecture and put forward an approach for evaluating workflow trust level using hidden Markov models and provenance data. We discussed how the HMM assumptions can be applied to this problem, and we provided details on how the model can be assessed using the provenance data and maximum likelihood method.

In order to investigate the behavior of the model, we provided a workflow scenario and expressed how its trust level is evaluated using the proposed model. Furthermore, we presented how the Viterbi algorithm was used to verify the HMM. In order to verify the effect of the stationary assumption of HMMs for the trust evaluation problem, we investigated the results of applying the non-stationary hidden Markov model to our problem.

The two models were then compared with each other. It was observed that the same trust level was estimated by both models with a small difference in their probability values. Therefore, the stationary assumption does not have a significant impact on the trust evaluation results. The non-stationary assumption of transition probabilities seems to be more accurate in case of our model since the probability of moving from one state to the other at a time instance depends on the state of the two services that are being executed at those times. Thus, for this problem, it is better to consider the transition probabilities as time-dependent probabilities for more accurate results.

Future work involves performing a large number of experiments to evaluate the scalability and accuracy of the system, preferably with real data. Various experiments will be done for different workflow sizes, and the behavior of the system will be observed in response to larger workflows.

As the amount of provenance data affects the accuracy of the learnt probabilities, the reliability of the system will be evaluated considering different learning data. We will also consider incomplete data and experiments will be performed with EM learning to estimate the results in case of missing data.

The main concern of the current implementation was randomly generating a large amount of valid provenance data for many workflows, each having some common pattern with others. The future workflows ought to be realistic and consist of common services and patterns with reasonable provenance values and data from a number of executions. The model will be improved to also consider trust values of the workflow process and input data for the evaluations.

Furthermore, the fluctuation of trust with the Markov process needs to be investigated in order to discover the points at which the workflow lacks trustworthiness and should be refined. It is desired to automatically detect and replace less trustworthy services with trustworthy ones. This part of the work will be extended by learning the workflow patterns from the provenance data and substituting less trustful services or sections of the workflow with more trustworthy ones.

References

1. Brown, B., Aaron, M.: The politics of nature. In: Smith, J. (ed.) The Rise of Modern Genomics, 3rd edn. Wiley, New York (2001)
2. Naseri, M., Ludwig, S.A.: A Multi-Functional Architecture Addressing Workflow and Service Challenges Using Provenance Data. In: Proceedings of Workshop for Ph.D. Students in Information and Knowledge Management (PIKM) in Conjunction with the 19th ACM Conference on Information and Knowledge Management (CIKM), Toronto, Canada (2010)
3. Gaaloul, W., Baïna, K., Godart, C.: Towards Mining Structural Workflow Patterns. In: Andersen, K.V., Debenham, J., Wagner, R. (eds.) DEXA 2005. LNCS, vol. 3588, pp. 24–33. Springer, Heidelberg (2005)
4. Altintas, A.: Lifecycle of Scientific Workflows and Their Provenance: A Usage Perspective. In: Proceeding of 2008 IEEE Congress on Services (2008)
5. Kim, J., et al.: Provenance trails in the Wings-Pegasus system. Concurrency and Computation: Practice and Experience 20 (2007)
6. Aiello, R.: Workflow Performance Evaluation. PhD Thesis, University of Salerno, Italy (2004)
7. Gil, Y.: Workflow Composition: Semantic Representations for Flexible Automation. In: Workflows for e-Science, pp. 244–257 (2007)
8. Rabiner, L.R.: A Tutorial on Hidden Markov Models and Selected Applications in Speech Recognition. Proceedings of the IEEE, 257–286 (1989)
9. Krogh, A., Mian, S.I., Haussler, D.: A Hidden Markov Model that finds genes in E. coli DNA. Nucleic Acids Research 22, 4768–4778 (1994)
10. Jelinek, F.: Self-organized Language Modeling for Speech Recognition, IBM T.J. Watson Research Center Technical Report (1985)
11. Bongkee, S., Jin, H.K.: Nonstationary Hidden Markov Model. Signal Processing 46, 31–46 (1995)

12. JingHui, X., BingQuan, L., XiaLong, W.: Principles of Non-stationary Hidden Markov Model and its Applications to Sequence Labeling Task. In: Proceedings of the Second International Joint Conference on Natural Language Processing (2005)
13. Fine, S., Singer, Y., Tishby, N.: The Hierarchical Hidden Markov Model: Analysis and Applications. Machine Learning 32, 41–62 (1998)
14. Altintas, I.: Lifecycle of Scientific Workflows and their Provenance: A Usage Perspective. In: IEEE Congress on Services 2008- Part I (2008)
15. Verdonck, F., Jaworska, J., Thas, O., Vanrolleghem, P.: Determining Environmental Standards using Bootstrapping, Bayesian and Maximum Likelihood Techniques: A Comparative Study. Analytica Chimica Acta 446, 429–438 (2001)
16. Fayyad, M., Piatetsky-Shapiro, G., Smyth, P.: From Data Mining to Knowledge Discovery: An Overview. In: Advances in Knowledge Discovery and Data Mining, pp. 1–34. AAAI Press / The MIT Press, Menlo Park (1996)
17. Forney, G.D.: The Viterbi Algorithm. Proceedings of the IEEE 61(3) (1973)
18. MySQL DataBase Software, www.mysql.com
19. Rajbhandari, S., Wootten, I., Shaikh Ali, A., Rana, O.F.: Evaluating Provenance-based Trust for Scientific Workflows. In: Proceedings of the Sixth IEEE International Symposium on Cluster Computing and the Grid (2006)
20. Rajbhandari, S., Rana, O.F., Wootten, I.: A Fuzzy Model for Calculating Workflow Trust using Provenance Data. In: Proceedings of the 15th ACM Mardi Gras Conference (2008)

Chapter 3
Unmanaged Workflows: Their Provenance and Use

Mehmet S. Aktas, Beth Plale, David Leake, and Nirmal K. Mukhi

Abstract. Provenance of scientific data will play an increasingly critical role as scientists are encouraged by funding agencies and grand challenge problems to share and preserve scientific data. But it is foolhardy to believe that all human processes, particularly as varied as the scientific discovery process, will be fully automated by a workflow system. Consequently, provenance capture has to be thought of as a problem applied to both human and automated processes. The *unmanaged workflow* is the full human-driven activity, encompassing tasks whose execution is automated by an orchestration tool, and tasks that are done outside an orchestration tool. In this chapter we discuss the implications of the unmanaged workflow as it affects provenance capture, representation, and use. Illustrations of capture include multiple experiences with unmanaged capture using the Karma tool. Illustrations of use include defining workflows by suggesting additions to workflow designs under construction, reconstructing process traces, and using analysis tools to assess provenance quality.

Keywords: Data provenance, e-Science workflows, provenance capture, data mining, case-based reasoning, intelligent user interfaces

Mehmet S. Aktas
Data to Insight Center, Indiana University, USA
e-mail: `maktas@cs.indiana.edu`

Beth Plale
School of Informatics and Computing, Indiana University, USA
Data to Insight Center, Indiana University, USA
e-mail: `plale@cs.indiana.edu`

David Leake
School of Informatics and Computing, Indiana University, USA
e-mail: `leake@cs.indiana.edu`

Nirmal K. Mukhi
IBM T. J. Watson Research Center, USA
e-mail: `nmukhi@us.ibm.com`

Q. Liu et al. (Eds.): Data Provenance and Data Management in eScience, SCI 426, pp. 59–81.
springerlink.com

3.1 Introduction

The data products produced during the course of workflow-driven scientific discovery have the potential to advance scholarly research and address pressing societal problems now and in the future. Nevertheless, however effective workflow systems have shown themselves to be at solving problems, there remain scientific discovery processes not amenable to representation within, and execution by, a workflow system. Workflows are inherently human processes, and it would be foolhardy to believe that human processes, particularly as varied as the scientific discovery process, can be fully automated. Computer scientists cannot hope to engineer the human out of the loop, nor can a workflow system promise to support within a single environment every tool scientists will ever use through the course of their research. Acknowledging this fact, we make the distinction between workflows that are executed end-to-end and fully under the control of a workflow orchestration system and those that are not, the latter we call the unmanaged workflow. The *unmanaged workflow* is the full human activity, encompassing tasks whose execution is automated by an orchestration tool, and tasks that are done outside an orchestration tool.

The issue of relevance to us with unmanaged workflows is provenance capture. An unmanaged workflow has a simple interpretation as two disjoint subworkflows with a gap between. We may know only that subworkflow-1, which began at time t_0 and completed at time t_i, occurred before subworkflow-2 which began at t_j and ended at t_n. Nothing more might be known about the relationship between the two. Given a distributed system with unsynchronized clocks, even this temporal relationship may not be known. The human activity occurring between the first workflow subworkflow-1 and second workflow subworkflow-2 could be the act of analyzing a result using a statistical package, could be the creation of a new layered product in at GIS tool, or could be simply a music break completely unrelated to either subworkflow-1 or subworkflow-2. Figure 3.1 illustrates this case.

For the human-in-the-loop workflow illustrated in Figure 3.1, the provenance of the humans actions may contribute to the provenance record of data product R. How can we know what human activity occurred between two workflow fragments? The two could be completely unrelated, and just mark a music and coffee break between two distinct and unrelated tasks. Or do we even need to know? The provenance of a piece of art has gaps in it; gaps that occur when the owner desired anonymity or when theft occurs. It may be sufficient to merely suggest that the two subgraphs are related and leave it at that. But suppose we can obtain provenance information from the human activity piece, how then can we stitch together the provenance from the human-action piece with the two subworkflows?

Mukhi [6] studies business workflow that cannot be fully automated, and addresses incompleteness through the notion of the unmanaged business process, which is a process that encompasses a large number of human driven workflows, the use of collaborative platforms to accomplish shared tasks, and handling of exceptional situations that arise in the context of automated workflows. BPEL4People [40] is an extension to WS-BPEL that allows people to participate in a business process. BPEL4People, though, requires a plan of activity (or a workflow) be known in

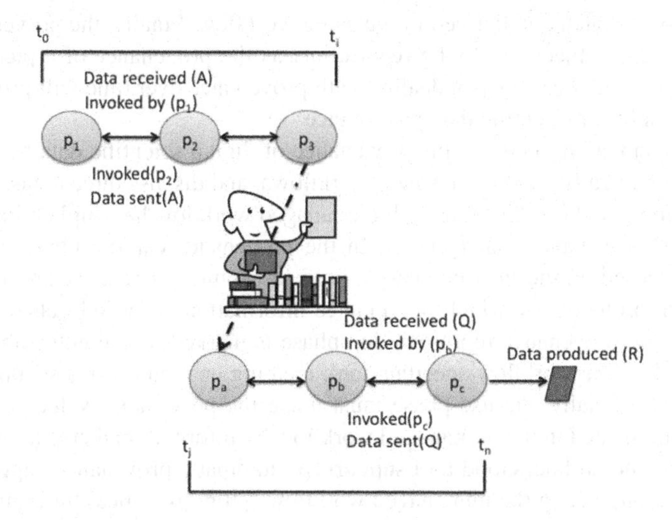

Fig. 3.1 Illustration of an unmanaged workflow

advance that describes the entire business process. BPEL4People, while useful in a limited sense, puts e-Science back at step one in that if scientists do not acknowledge that what they are doing can be described by a workflow, how can one possibly be specified in advance?

The unmanaged workflow defines a problem space of provenance capture wherein two things occur, first, there are non-automated steps in the activity, and second, the full activity cannot be specified in advance. This kind of workflow which is prevalent in e-Science though not in business where workflows are better understood, is a grand challenge for provenance capture.

With no single workflow specification to guide provenance capture, the effects are cascading. Provenance capture becomes more difficult because there is no guidebook of what is supposed to happen, nor is there a single workflow orchestrator that controls execution and determines failure and execution models of the workflow. Representation of the provenance in a provenance store has to deal with fragments of provenance because the captured provenance has a higher likelihood of being ad hoc, noisier, and less complete. Finally, the use of the provenance, though having much in common with use of the provenance of managed workflows, has unique challenges because of the ad hoc nature of the information.

This latter point of the ad hoc nature of provenance has significant implications for trust. A key benefit of having the provenance record of a scientific data object or set of objects is that someone with whom the data is shared can use the provenance to determine their level of trust in the data. If the provenance itself is of questionable quality, it undermines one of the key benefits of provenance of scientific data in the first place. On the other hand, thinking about the provenance record in terms of fragments of workflows that are either part of the lineage trace of a scientific data object or not related, models reality more closely. This is mainly because data are rarely created from scratch and derived in the same workflow. Finally, the provenance of a scientific data object is a living record, just as the provenance of a piece of art is a living record. Flexibility in dealing with provenance over time will provide the greatest benefit for scientific data provenance.

In this chapter we focus on the provenance of digital scientific data that is generated automatically from unmanaged workflows, and discuss three areas, capture, representation, and use. Dealing with unmanaged workflow has implications at every step of provenance management. In the provenance capture phase, attention must be focused on the instrumentation used to capture provenance and the communication protocols by which provenance information is ferried outside the application. The provenance representation phase is marked by the noted absence of a overall plan (the workflow specification), creating uncertainty as execution-level events arrive. Finally, the use phase must make the provenance valuable for use. This is complicated in the unmanaged workflow by information that is known to be incomplete and ad hoc. Good tool support for automatic provenance capture, representation, and use in the unmanaged workflow setting is critical for realizing the vision of broad scientific data sharing today and in decades to come.

3.2 Provenance Creation

3.2.1 Overview

Provenance creation for unmanaged workflow is the activity of identifying the important provenance activities, defining a data model by which to represent the provenance, mapping the activities to the model, identifying the right communication protocols and instrumentation techniques to employ, then finally, putting capture in place for running applications. A good graph-based model for provenance is the Open Provenance Model (OPM) [7]. OPM represents entities, artifacts, actors, and relationships in the form of a directed graph. The provenance of a data object, D_i, for instance, can be defined by the processes or transformations that were applied to create the data object. The processes can in turn be further described by their inputs and outputs. A relationship between process P_1 and process P_2 exists if process P_2 consumes a data product generated by process P_1. OPM defines the minimal provenance but supports name-value pair annotations that can be used to enhance the information known about the entities and relationships. Provenance can be further enhanced by extending the existing set of relationships and objects, such as was done by Missier et al. [22].

In the unmanaged workflow setting, extracting provenance from an executing application has similarities to real time performance monitoring of a complex, distributed and parallel application. The terminology used in performance monitoring literature when referring to the mechanism for extracting information from an executing application [25] is "instrumentation", "instrumentation points", and "sensors", so we adopt the same terminology here. Provenance capture focuses on the sensors that collect information; as with performance monitoring these sensors must be lightweight, and minimally perturb or pollute the application.

One of the first design decisions is the types of instrumentation that can be supported. Mechanisms for collecting provenance have tradeoffs that must be made between burdening the user, the developer, or application performance; and in the ultimate quality of the provenance information as well. Collection mechanisms fall into one of three categories: user annotation, scavenging, or full provenance instrumentation. Provenance capture through *User annotation* is a human data entry activity where users enter textual annotations, capturing for instance all the data sets used during an analysis of a particular forest use in the Amazon forest over a multiyear period, including video interviews of nearby residents, satellite imagery, and survey data. To make the entry more uniform, the scientist might be prompted to enter specific information. It is however widely understood that user-entered metadata is often incomplete and inconsistent [26]. The annotation approach imposes a low burden on the application, but a high burden on the humans responsible for annotation. The implication is that error rates of the provenance are high.

Full provenance instrumentation refers to instrumentation that is added directly to an application, such as when a programmer must insert calls into their code to call out to a provenance library. Full provenance instrumentation allows for good provenance completeness and consistency, but imposes a substantial burden on the programmer who must modify the application directly. A middle approach is something we refer to as *scavenging*. Here collection is done by means of piggybacking onto existing collection mechanisms, such as a logging tool or an auditing tool, or is carried out in the middleware layer so as to not burden the application programmer. VisTrails [27] implements a form of scavenging when it captures the "do" and "undo" actions of graphical modeling tools as a way to pick up provenance for free. Scavenging has a disadvantage of resulting in incomplete information. Incomplete provenance information can be an acceptable tradeoff for high levels of collection interoperability as long as we can provide a sense of the level of completeness and provide an estimate of the accuracy of the resulting provenance.

The capture layer ingests provenance events, alternately called "notifications", as they are generated at runtime, and queues them for storage to a provenance capture system. The layer is implemented as a protocol and framework to carry provenance events from application components to the database. In this layer, there may be different protocols ranging from a Web Service based system to a publish-subscribe system. This is illustrated on the left of Figure 3.2.

3.2.2 *Application in the Karma Tool*

The application of provenance capture in unmanaged workflows is best illustrated through example. Our team has had experience applying provenance capture in multiple and varied settings, in which we had to think through the data model, instrumentation mechanisms, protocols. We summarize this experience in Table 3.1. PC3 is Provenance Challenge 3 [37], a friendly competition undertaken June 2009 in the

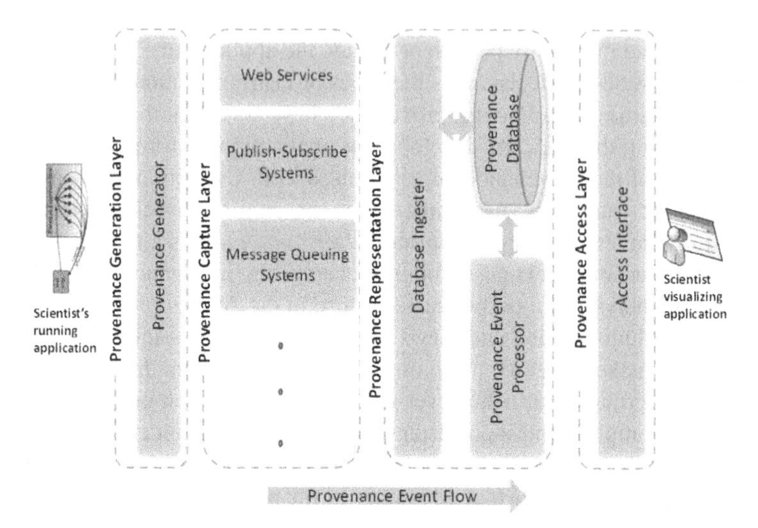

Fig. 3.2 Logical architecture of a provenance system

provenance community. Teams implemented an astronomy workflow that used their provenance system to answer a set of queries. Ratings depended on the number of queries the systems could answer and the ability to capture the most information in the query. AMSR-E is shorthand for a satellite instrument processing pipeline application. The processing pipeline is for the Advanced Microwave Scanning Radiometer - Earth Observing System sensor located on the NASA Aqua satellite. The pipeline is a script-driven application that continuously ingests images from the polar orbiting satellite, processes the images to identify sea ice over the poles, etc. The pipeline is made up of legacy processing scripts and scientific algorithms, the latter of which are of deep importance to the provenance record. The LEAD application is a workflow system for executing weather related analysis and modeling activities. The GENI application is a computer networking application. It applies provenance capture to the PlanetLab distributed network. Table 3.1 identifies three different kinds of instrumentation and two protocols that are used in the four applications, in some applications two instrumentation techniques are used. The implications of the capture mechanism, using the vernacular identified in Table 3.1, are given in the column titled "Provenance capture burden". As it can be seen from Table 3.1, there are various instrumentation mechanisms, one of which is a full provenance instrumentation approach, with the attendant high burden on the application programmer to correctly place the instrument points in the code, and to write provenance events to a format that Karma requires. We see this as our least viable solution because of this dual-headed burden. The first scavenging approach taps into the messages flowing between applications and transparently routes a copy of the event to Karma. It treats the application as a black box, so provenance is limited to what can be captured through message traffic. Within that approach, there are two schools of thought as to whether or not one should peek into the message contents to extract further provenance. The second scavenging approach captures provenance information from log files.

The Karma system currently supports two forms of communication in which the provenance event come to the system, a Web service based model and a publish-subscribe messaging system [38, 39]. In addition to synchronous submission of notifications, support for asynchronous publishing of provenance is supported through a publish subscribe system. Sometimes called an Enterprise Service Bus, a publish subscribe system decouples publishers of events and consumers of events, allowing new publishers and subscribers to join simply by having the topic name and location of the broker that brokers subscriptions.

3.3 Provenance Representation

A provenance system can be viewed as a repository that (1) actively collects and ingests events in real time, (2) stores the events in a data model that supports time-series data storage, aggregation and synthesis of the events to form new knowledge, and (3) provides an access layer that supports access to the data. Provenance systems are often designed to serve a single use, such as provenance capture for a single workflow system. As attention is increasingly being paid to the long term sharing and preservation of digital scientific data, provenance systems can be valuable repositories of information about the circumstances under which data objects were created, information that is essential to reuse of the data object in a new setting.

The provenance representation layer, illustrated in the middle of Figure 3.2, stores provenance data using a data model that represents the execution instance notifications, and higher layers that abstract from execution instances. The representation layer is where post-processing is carried out such as to organize the events and derive higher levels of behavior or knowledge from the events. The representation layer is where the impact of unmanaged workflows is most strongly felt because for unmanaged workflows there is no obvious reference point to which arriving execution events can be tied such as would be provided by a workflow known in advance.

Provenance systems use different data models, Karma uses a two level model; Trident and VisTrails use a three level data model. Karma includes both execution details for utilizing the data and high level information for long term preservation [41]. This layer should contain information about services and data products at a sufficient level of detail to support discovery and automated decisions about whether to bind a particular data product or service. This layer should contain information for locating and retrieving data artifacts for use in a workflow execution and capture instance invocation and execution details of a particular run. The representation layer should store common information consistently and without redundancy. A provenance capture system captures provenance by accumulating discrete run time activities during the lifecycle of unmanaged workflows, that is, workflows whose structure is not known to the system in advance of execution.

Table 3.1 Instrumentation techniques supported by Karma and the two communication protocols have different tradeoffs [44].

Instrument mechanism	Communication protocol	Provenance collection burden	Application
Application responsible for invoking library that constructs provenance notifies	and invokes Axis2 send call. Specialized Axis2 handler routes message to Karma.	Programmer must be provenance savvy. High application burden.	PC3
Application responsible for invoking library that constructs XML provenance notifies	and publishes to messaging system (i.e. RabbitMQ). Karma listens for events.	Application programmer must be provenance savvy. High application burden.	AMSR-E
Application publishes SOAP notifications as part of normal activity	and publishes to Axis2 call. Axis2 handler transparently grabs copy of event and sends to Karma without application being aware.	Karma parses notifications on server side to extract useful provenance information. Assumes basic provenance behavior is present in message. Scavenging approach.	LEAD
Application publishes notifications as part of normal activity	and publishes to RabbitMQ. Karma is sitting on topic/-subject so captures event without application being aware.	Same as above. Scavenging approach.	GENI
Application writes log messages to log file as part of normal activity	and Karma Adaptor parses log file (client side parsing) and generates notifications that are sent via Axis2 or RabbitMQ.	Adaptors need to be written to parse log file; assumes core provenance behavior has been written to log. Scavenging approach.	GENI, AMSR-E

3.3.1 Representation in Karma

Karma stores provenance data using a two-layer information model which includes both execution details for utilizing the data and registry information for long term preservation [41]. The two-layer information model contains a registry level, which contains metadata about the instance, and an execution level. The registry level has similarities to registries used in web service architectures in that it contains information about services and data products at a sufficient level of detail to support discovery and automated decisions about whether or not to bind a particular data product or service. The registry level is not used for locating and retrieving data artifacts for use in a workflow execution, but nevertheless contains sufficient information for building a data object that can be preserved indefinitely. The execution level captures instance invocation and execution details of a particular run. The two-layer model recognizes commonalities in workflows and stores that common information consistently and without redundancy. Some of the concepts of this two-layer information model map directly to OPM. For instance, data products such as data granule and data collection can be considered artifacts; entities (services including composite service and opaque service, and methods) can be considered processes; and clients (a kind of entity, which may be a user or a workflow engine that initiates the workflow) can be considered agents.

A significant implication of not knowing the structure of workflows in advance is that in addition to not having a picture of execution before it occurs, Karma can make no assumptions as to the existence of global state in the application either. A provenance notification message will be issued by a task, and the information contained in that notification will be based on what can be gathered from local state only. For instance, tasks within a workflow may not know the session or workflow to which they "belong" so it can be difficult for the Karma service to tie a service invocation back to the workflow that invoked it, particularly for recursive or chained services. For unmanaged workflows, OPM is inadequate to the task of defining the formats of provenance events if for no other reason than provenance capture is messier and more incomplete than the graph-based OPM can handle.

3.4 Provenance Use

Captured provenance provides a rich source of information about workflow execution. Automatic provenance collection over time generates a substantial body of knowledge which may be used in many ways. Referring back to the logical architecture shown in Figure 3.2, the access layer supports the Query API, which is used to pose queries to the provenance capture system to retrieve provenance. This layer that allows users to explore and examine large quantities of data requires browse capabilities. The browse pattern characteristically involves starting with some broad information, performing a search, finding general result sets and then selecting more specific information for drill-down. The access layer includes macro-level queries, tracing provenance relationships back through time in order to construct a graph, as well as object-level queries that locate information about specific entities matching the conditions specified in passed arguments.

Our research investigates three novel additional uses for provenance, described in the following sections: Using captured provenance to aid workflow construction, to repair problems in provenance capture, and to analyze workflow trace quality.

3.4.1 Using Provenance to Aid Workflow Construction

Provenance acquired by provenance capture systems is rich source of information about the workflows which gave rise to the observed processes. We are investigating how such information can be used to aid workflow construction. When a subpart of a partially constructed workflow involves a sequence of services observed in a past provenance instance, that remainder of the stored provenance can suggest extensions of the partial workflow, to present to the workflow author. We are exploring both the mining of stored provenance for statistical correlations on which to base predictions, and the use of case-based reasoning (CBR [48, 49]) to predicting solutions to new problems based on relevant instances of similar prior problems. CBR is a "lazy learning" method in that cases are stored with minimal pre-processing, simplifying knowledge acquisition. Because CBR reasons from relevant prior episodes—cases—rather than rules, it is a natural approach for reasoning from libraries of examples such as provenance databases. The performance of CBR systems depends on how well their stored cases cover the space of problems to solve. Large-scale provenance databases provide an extensive starting point, and each new workflow execution provides a new case to extend coverage. In addition, workflow problem types tend to recur—for example, scientists in a particular domain will tend to generate certain types of workflows [50], increasing the chance that stored traces will be relevant to new situations.

The Phala project[1] [50, 51] develops and tests a case-based approach to aiding workflow construction. Phala is a plug-in to the XBaya graphical workflow composer [45]. Phala's processing cycle is illustrated in Figure 3.3.

As a user develops a workflow, Phala monitors the partially constructed workflow and generates background retrieval queries to a provenance database. The provenance database need not have been generated for a single particular task; similarity assessment process selects those cases which are relevant. When cases are not available, the system can provide recommendations based on statistical methods, which use statistics mined from the provenance database. Because there is no guarantee that the suggestions generated by statistical methods and cases will agree (or even that all suggestions from relevant cases will agree), we are developing approaches

[1] The name Phala was inspired by the naming of the Karma provenance capture system. In Sanskrit, Karma means causality and reflects captured provenance, Karma means the ripened fruit, so KarmaPhala means the fruit of provenance capture, reflecting the cases generated by Phala.

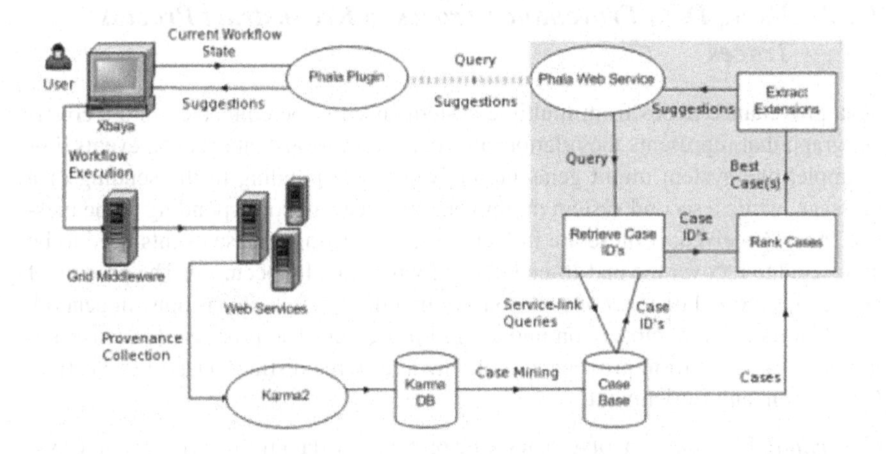

Fig. 3.3 Phala's processing cycle

for combining multiple (and possibly conflicting) recommendations, in order to extend the range of situations for which Phala can make recommendations and increase recommendation accuracy. Initial tests of these methods are promising [51]. In addition, Phala allows users to control the level of confidence required for the system to propose suggestions.

Large provenance case bases provide both benefits and challenges for CBR. Given a large and diverse set of cases, a CBR system can solve a wide range of problems. However, retrieval and similarity assessment for large case bases can be computationally expensive, especially when the cases which need to be compared involve structured information. For graph-structured case information such as workflow traces—for which matching could be seen as an instance of the subgraph isomorphism problem—matching cost is a particularly acute issue. In addition, the anticipated size of provenance case bases far exceeds that of case bases previously studied by the case-based reasoning community (for example, as described in Section 3.4.3, a 10 GB database was recently developed as a provenance testbed). Consequently, a central goal of the Phala project has been to develop procedures enabling efficient retrieval of structured cases.

For generality, Phala's retrieval methods are primarily domain-independent. Phala's retrieval is performed by the Structure Access Interface (SAI), a toolkit for structure-based retrieval. To increase retrieval efficiency, SAI implements a two-phase retrieval approach in which the initial phase can be seen as coarse-grained filtering, to retrieve a small set of potentially relevant cases for more expensive structural matching. More detailed descriptions of the algorithms and evaluations are omitted here for reasons of space, but are available elsewhere [50, 51, 53].

3.4.2 Using Data Provenance Traces to Reconstruct Process Traces

Data provenance traces from multiple systems need to be connected into a coherent graph that represents the relationships between various data-related events. For example, one system might generate an event corresponding to the sending of a message, while a second system might generate an event corresponding to the message being received: if these are in fact the same message, these events need to be connected to recover the end-to-end trace of what actually occurred. The process of doing so is termed trace reconstruction; we first discuss how this is done in general, and then focus more closely on unmanaged processes. The process of reconstructing process traces from provenance data involves three distinct phases: Collection, Correlation, and Enrichment.

Collection: This phase involves gathering provenance data from various source systems. Adapters are built to extract events or log information from the source system, perform appropriate transformations to produce provenance items and then record these provenance items into a centralized provenance store. For reconstructing process traces, a provenance solution would involve deploying adapters to all systems where any process activity occurs, such as document repositories, web servers, email servers and so on.

Correlation: This phase involves correlation of provenance items within the provenance system. Correlation for the purpose of reconstructing a process trace will involve using an opaque process identifier if available, or a set of application data that collectively serves as the identifier for a process, and then using the identifier to stitch together the tasks, data and actors involved in the correct temporal sequence. The correlation will also locate identifiers that help to bridge systems (such as a message identifier that helps us connect a message sent from one system to that received in a different system).

Enrichment: When reconstructing process traces, provenance items recorded as multiple tasks by adapters may together correspond to a single process activity from the user's perspective. Creation of such a higher level abstraction would be done at this time. For example the entire sub-graph showing the sending of a message from one system to another could be abstracted into a single node labeled 'message exchange', with the underlying provenance of this activity preserved elsewhere in the graph. This simplification serves two purposes: it allows for easier visualization and also reduces the complexity of consuming new information. For example, further correlation or enrichment of the graph could be triggered on new 'message exchange' node, rather than on the more complex pattern it corresponds to.

It is important to note that these phases need to operate concurrently; data that is being recorded has to be correlated and enriched at the same time other data is arriving; i.e. development of the provenance information and its use to reconstruct the process trace is a continuous process. The outcome of this continuous process is a process trace represented as a provenance graph.

A result of the observed process being unmanaged is that sometimes correlation and enrichment are non-trivial. Consider the process of running a long-running scientific experiment across a computing infrastructure that requires data products from one system to be manually copied to a different system. In many real-world instances, the method for achieving this is for scientists running the experiment to request system administrator to perform the data copy. The request forms an important part of the overall process and allows the subsequent data set used to be verified as being the correct one. However, correlating an unstructured request such as an email with other events such as a file copy is non-trivial. In most cases, the request would specify the source file and target location, but this may not always be the case. In such instances, the time the request was made could be used for correlation. In general, it is certainly possible to miss the request that was used when creating the provenance graph, or to correlate an incorrect request. We classify uncertainties in provenance graphs under three categories: Node versus Edge uncertainty, Simple versus Complex Uncertainty, and Static versus Dynamic uncertainty.

Node versus Edge uncertainty: All the nodes in the provenance graph can be grouped into two categories, nodes recorded by adapters (Type A) and nodes created through derivation from those (Type B nodes). Nodes recorded by adapters are accurate since they exactly represent something that occurred in a source system. Derived nodes, added through feature extraction or enrichment may however be inaccurate, and may therefore have uncertainty associated with them. Edges are recorded based on correlations or shared features between nodes. When edges are recorded between Type A nodes that were recorded from the same source system, they are guaranteed to be accurate (since they are based on correlations of consistent and accurate data). All other edges may be imprecise. For example, an edge showing time ordering between Type A nodes from different source systems may be inaccurate if the clocks are not synchronized. Additionally, edges between Type B nodes may be imprecise since the node uncertainty is propagated to the edge, i.e. the edge may be created based on imprecise data. Going back to our earlier example of correlating a request to copy data with the actual system activity corresponding to the data movement, the resulting provenance graph would have accurate information on each of these basic events, the uncertainty would arise when trying to determine which of the available email requests correspond to a particular data movement activity.

Simple versus Complex Uncertainty: Uncertainty associated with a provenance item may be entirely a function of features of that provenance item itself. We call such uncertainty simple. Sometimes the uncertainty is a function of a set of provenance data; in such cases it is said to be complex.

Static versus Dynamic uncertainty: When the uncertainty associated with a given provenance item is fixed, it is said to be static uncertainty. Sometimes the uncertainty associated with a piece of information varies over time. This is dynamic uncertainty.

Doganata et al. [43] have explored an approach in which information is correlated only if it appears to meet a certain confidence threshold. We [6] have found that it is better to represent uncertainty in a first class manner within the provenance graph itself, and allow for provenance applications or further enrichment to decide what information is accurate, given a more global view of the end-to-end process.

3.4.3 Using Provenance for Analysis of Workflow Traces

Real world provenance data traces are often noisy, resulting in disjoint or incomplete provenance traces. Provenance messages may be dropped, messages can be incomplete (which could occur when the application scope at a point of notification generation is more restricted than anticipated), or execution of the application (or workflow) can simply fail. To properly asses captured workflow traces, it is crucial to establish algorithms to analyze those traces, identifying failures and assessing the quality of data provenance traces.

Identifying failure in provenance traces: We have developed a model for analyzing provenance traces and identifying failures [52]. The model identifies two types of failures: a) task failures where a node in a workflow does not complete successfully, b) communication failures in which a task completes but the notification is not successfully transmitted.

We have performed experiments studying four failure modes as follows: a) No failures and dropped notifications (success case), b) 1% failure rate, c) 1% dropped notification rate, d) 1% failure rate and 1% dropped notification rate. These failure rates are modeled using uniform distributions in a workflow emulator, WORKEM [54], to determine if a particular invocation must fail or drop a notification. Using the WORKEM to generate provenance, the following six major workflows were used as the basis for generating a large scale (10 GB) provenance database: LEAD North American Mesoscale (NAM) initialized forecast workflow, SCOOP ADCIRC Workflow, NCFS Workflow, Gene2Life Workflow, Animation Workflow, MotifNetwork Workflow. These workflows are pseudo-realistic, in the sense that they are modeled after real life workflows. The LEAD NAM, SCOOP and NCFS are weather and ocean modeling workflows, Gene2Life and MOTIF are bioinformatics and biomedical workflows, and the Animation workflow carries out computer animation rendering. Some of the workflows are small, having few nodes and edges, while others like Motif have a few hundred nodes and edges.

Figure 3.4 shows the results where the distribution is dissimilar. Even though the generation settings for WORKEM were identical across workflows, WORKEMs failure model does not result in the same uniform distribution across different workflows since the configuration for failure rates is per task in the workflow. For both Animation and Motif workflows, the number of runs that do not have failures or dropped messages is approximately half of what the smaller workflows exhibit, which supports that the larger a workflow, the higher the failure rate and dropped messages rate. The smaller workflows appear to have the same distribution compared to each other.

Quality Assessment of provenance traces: We have developed a methodology for assessment of the provenance goodness [52]. This methodology applies statistical approaches that operate over large volumes of data to zero in on suspicious

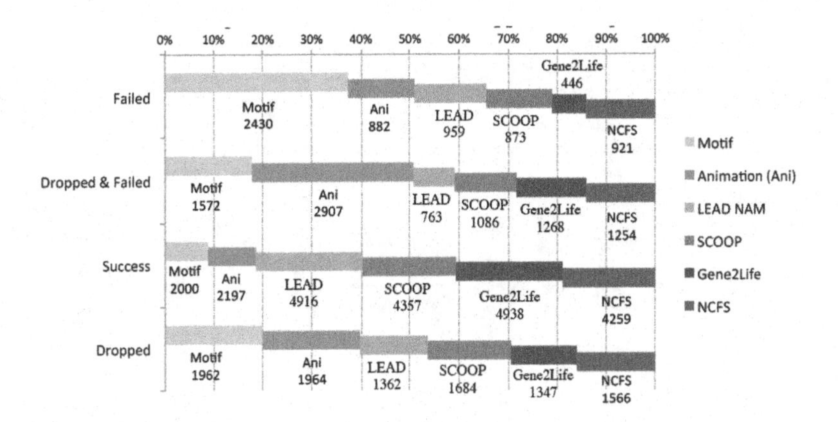

Fig. 3.4 Distribution of workflows by population cases

provenance records. Based on this approach, provenance goodness is determined by constructing the best possible provenance graph for an execution based on the captured provenance record, then assessing the goodness of the resulting graph by looking at the partitions in a provenance graph. In this approach, a provenance graph is be modeled as PG = V, E, where V is a collection of vertices that are linked by one or more directed edges, E. This approach is used to construct a provenance graph from nothing (no guiding workflow template) based only on the captured provenance. It relies on an assumption that all provenance notifications contain the correct ID for the workflow execution instance to which they belong. WORKEM workflow emulator supports this assumption. Even with this simplification, this approach still may yield disconnected components. The query of a graph using a workflow ID searches over the database tables for entities (processes) that have matching IDs. If there are dropped messages, the queried graph may have missing edges or missing vertices. The only guarantee for the retrieved graphs is that the components of the graph are linked through that workflow ID.

Figure 3.5 shows the results of this approach, when the algorithm is applied to the aforementioned large scale 10GB provenance database. Observing the number of edge counts of each workflow instance, we conclude that the results for the LEAD NAM workflow are preliminary. The plot points are classified based on the statuses of each workflow. Based on the results, the workflows with dropped messages cluster towards the upper end of Figure 5. This implies that dropped messages for successful workflows are few. In comparison, workflows that involve failures typically result in more missing notifications, resulting in lesser number of edges in their provenance graphs.

Automatic provenance repair: As described previously, automatic provenance capture is imperfect. Correlation and enrichment methods aid in reconciling and merging information, but in some cases, the messages relating to a process may simply be lost, resulting in gaps in the provenance trace. Consequently, we are studying

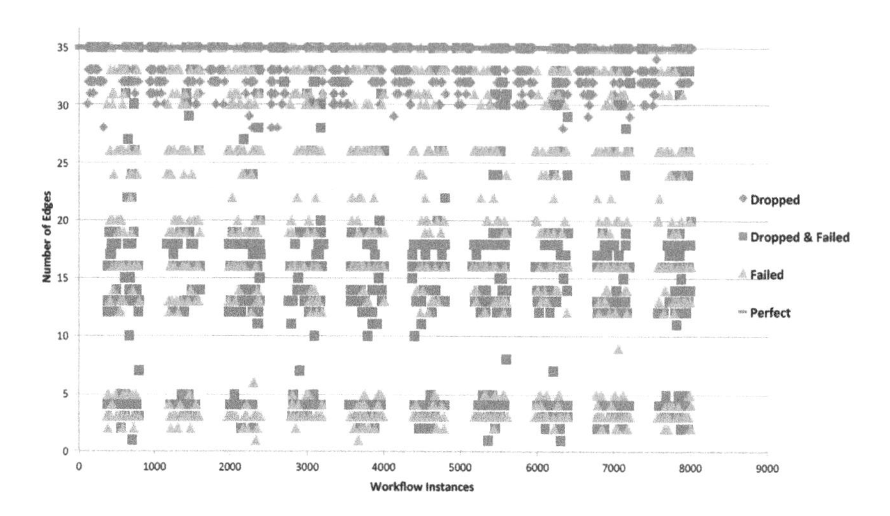

Fig. 3.5 Plot of edge counts for LEAD NAM workflow instances with different statuses

methods for assessing overall provenance goodness. For example, we have begun to explore using simple graph analysis methods to identify connected components: If a provenance graph is disconnected, it reveals gaps in provenance capture. In such instances, methods such as Phala's prediction methods, applied to the disconnected components, may be able to infer missing steps. Confidence assessment methods already developed for Phala's recommendations may be used to determine confidence in proposed repairs. This will enable a flexible repair approach in which repairs are only pursued if their confidence exceeds a use-designated threshold, and will also enable annotating repaired provenance with confidence values for the quality of the provenance.

3.5 Related Work

Provenance has been studied from different perspectives and several surveys have been published [4, 21, 28, 29]. Simmhan et al. [4] introduced a taxonomy of provenance in e-science, specifically for scientific workflow systems, based on why they record provenance, what they describe, how they represent and store provenance, and ways to disseminate it. Moreau [29] did a comprehensive survey analyzing 425 papers in the provenance literature and reviewing its potential benefits in e-science, curated databases, and semantic web. In this section, we give a brief review of provenance capture in major provenance systems particularly designed for scientific workflows. Generally, provenance systems for scientific workflows can be categorized into two classes: those in scientific workflow systems and standalone provenance systems.

Provenance Use: Provenance systems track artifacts from various systems and correlate these to create a provenance graph. Using a provenance graph, it is possible to recover process traces or data lineage, or to maintain a record of user activity for various purposes. It is also possible to build systems that aid workflow construction. Workflows are often generated by scientists who are not experts on scientific computation, who may have difficulty choosing appropriate services. Even for experts, workflow generation may be time-consuming. Consequently, software to facilitate workflow generation is highly desirable and a number of efforts have aimed to assist scientist in workflow generation, using both generative planning and interactive approaches. Systems such as FlowRecommender [11] and Viscomplete [12] mine process traces for the development of automated workflow recommendations. Wen et al. introduced an approach for process mining problem in dealing with invisible tasks, i.e., such tasks that exist in process model but not in its event logs [10], for deploying new business processes as well as auditing, analyzing and improving already interacted ones. Interfaces such as XBaya [45] aid users by abstracting away from the details of workflow languages. Knowledge-rich artificial intelligence methods have been developed to generate workflows automatically [46] and to provide interactive support for carefully codified domains [47]. Such approaches can provide excellent performance, but at the cost of expensive knowledge capture, which becomes a major impediment to fielding such systems in new domains. We are investigating data-driven methods to support human workflow generation with minimal knowledge capture.

Provenance tracking has been used to assist in reproducibility of scientific experiments [16], monitoring complex processes that span multiple systems [15] and measure compliance of unmanaged processes [17]. However, none of the existing literature has dealt with the the issue of uncertainty in provenance data. We are investigating the uncertainties involved in creating provenance traces for unmanaged processes and how to represent these using a provenance data model.

Systems such as Karma address provenance capture that is tightly coupled to a workflow system and provenance capture in non-structured e-Science environments. These systems provide a controlled provenance generation environment and do not necessarily contain provenance with failures.

A number of synthetic workflows have been generated and used in distributed systems [18, 19, 20] and computing networking [13, 14] research areas for performance evaluations and benchmarking purposes. However, none of these workloads attempt to model failures and have been specifically developed for the purpose of provenance research. As discussed previously, we use a noisy 10 GB provenance database that models failures of provenance notifications to explore methods to provenance repair and provenance quality assessment.

Provenance Systems: Provenance collection is widely supported in scientific workflow management systems because provenance data can be easily captured and recorded during execution of a workflow. Kepler [30] is used by scientists in multiple disciplines for the design and execution of workflows. The provenance component, Provenance Recorder (PR), is optional depending on the user's requirement to track provenance. To enable provenance capture in a workflow instance, the user drags the provenance recorder from the toolbox, places it on the workspace, and fills in the configuration menu. The provenance associated with the workflow definition is automatically generated by the existing MOML (Kepler's internal XML workflow representation) generation capabilities during a workflow run. To receive the provenance data, the PR implements several event listener interfaces. When the workflow is loaded, the PR will register with the appropriate concerns in the workflow. When the workflow is executed, PR will process information received as events, and save it in provenance store.

The Taverna workbench [31] is developed for the composition and execution of workflows for the life sciences community. Provenance data is recorded for workflows in the Simplified conceptual workflow language (Scufl) language with four levels [32]: a) process provenance, b) data provenance, c) organization provenance, and d) knowledge provenance. The process provenance records the order of service invocations, inputs/outputs to these services, and the time information of service invocations and workflow executions. The data provenance builds a derivation path of data objects in a workflow run. The organization provenance stores the metadata for the experiment such as who, when, and where the information was created and how it evolved during experiments. The first two levels of provenance are automatically logged during workflow execution. The organization and knowledge provenance can be obtained from three different sources: users' annotations of the Scufl workflows through a knowledge template plug-in; service descriptions from the myGrid semantic service discovery component Feta; and provenance published by the third-party data providers [33].

VisTrails [27] is a workflow and provenance management system that provides support for scientific data exploration and visualization. It is designed to handle rapidly-evolving workflows by using a change-based provenance model. The VisTrails provenance information is organized into three layers: workflow evolution, which captures the relationships among the series of workflows created in an exploratory task; workflow, which consists of individual workflows; and execution, which stores run-time information about the execution of workflow modules. The information for the first two layers is naturally captured by the change-based provenance mechanism. When a user modifies a workflow, his/her actions are captured by the History manager and saved in the VisTrails Repository. Run-time information is captured by the Workflow Execution Engine and stored in the VisTrails Log. Annotations are allowed at all levels of the layered provenance model.

The Trident [34] workbench is a scientific workflow system which is built on top of Windows Workflow Foundation (WF), a workflow enactment engine included in the Windows operating system. It provides an integrated way to collect, store, query, and view provenance for scientific workflows. Provenance information in Trident is a combination of the workflow schema—static, composition information about the workflow—and the provenance schema—dynamic, runtime information about the actual execution of a workflow instance. The Workflow Composer is the primary source of the workflow schema, and the Execution Service and Windows WF engine are the two main sources of the provenance schema. The Execution Service tracks the submission of each workflow instance, and the Windows WF engine natively generates tracking events of the workflow execution. Trident uses the BlackBoard [35] publish-subscribe, asynchronous messaging framework, to distribute the events from the source to the provenance storage. The Provenance Service listens to the events and records them in the provenance store. Trident event handlers listen for the built-in events to trace the workflow's control flow. The data flow knowledge obtained from the input and output parameter values passed to/from the activities are captured by the instrumentation in the Trident base activity class to generate customized user events.

The PASAO project provides an interoperable way to collect provenance in a grid environment using an open provenance protocol. Miles et al. [36] analyzed 23 use cases in biology, chemistry, physics, and computer science and determined 14 technical requirements for a generic, application-independent provenance architecture. PASOA is designed in three layers: fundamentals of recording and access, querying, and processing. PASAO supports the recording and use of three types of provenance: interaction provenance, which records interactions between components and data passed between them; actor provenance, which records processes information and the time of the execution; and input provenance, which records the set of input data to infer a data product. Groth and Moreau described the recording protocols in [9] in detail. Additionally, Frew et al. [23] captures application calls to the operating system (i.e., kernel calls) and Holland et al. [24] captures file system access.

3.6 Current and Future Challenges

The notion of unmanaged workflows reflects acceptance that human processes cannot be fully automated. For whatever reason, there are pieces of the workflow that remain with the user or are executed outside of and away from the "eyes" of the workflow system. If the task of provenance capture is to record a complete provenance or lineage record then the task is doomed to failure. This gives rise to the question: What can be done? The assumption in the phrase "complete provenance" must be revisited. Just as the provenance of a work of art may have gaps, so incompleteness in provenance of scientific data may be more common than we may think. Methods such as trace reconstruction may help to fill in the trace. However, is complete provenance necessary, or even desirable? Provenance capture can result in large volumes of very low level information. Provenance capture in the AMSR-E satellite imagery processing stream reveals significant amounts of "housekeeping" information, such as that the processing script ran on a certain day. Scientists who question a resulting image are interested in the version of the science algorithm that was applied, but not in the specific day. How can we separate the wheat from the chaff to identify the provenance that contributes meaningfully to the final outcome? Finally, too few examples of compelling uses of provenance captured from real applications exist to convince communities of users that provenance systems are worth the investment of time. These questions and challenges make provenance and provenance use a rich research area for the future.

Acknowledgements. We thank Bin Cao of Teradata Corp. and Yiming Sun, You-Wei Cheah, Yuan Luo and Peng Cheng of Indiana University for their contributions and discussions. This research funded in part by the National Science Foundation under grant OCI-6721674, by BBN Technologies, and by NASA under NNX10AM03G.

References

1. Droegemeier, K., Gannon, D., Reed, D., Plale, B., et al.: Service-oriented environments for dynamically interacting with mesoscale weather. Computing in Science and Engineering. IEEE Computer Society Press and American Institute of Physics 7(6), 12–29 (2005)
2. Folino, G., Forestiero, A., Papuzzo, G., Spezzano, G.: A grid portal for solving geoscience problems using distributed knowledge discovery services. Future Generation Computer Systems 26(1), 87–96 (2010)
3. Deelman, E., Gannon, D., Shields, M., Taylor, I.: Workflows and e-Science: An overview of workflow system features and capabilities. Future Generation Computer Systems 25(5), 528–540 (2009)
4. Simmhan, Y., Plale, B., Gannon, D.: A survey of data provenance in e-Science. ACM SIGMOD Record 34(3), 31–36 (2005)
5. Simmhan, Y., Plale, B., Gannon, D.: Towards a Quality Model for Effective Data Selection in Collaboratories. In: IEEE Workshop on Workflow and Data Flow for Scientific Applications (SciFlow 2006), Held in Conjunction with ICDE, Atlanta, GA (2006)
6. Mukhi, N.K.: Monitoring Unmanaged Business Processes. In: 18th Int'l Conf. on Cooperative Information Systems, CoopIS (2010)
7. Moreau, L., Clifford, B., Freire, J., Futrelle, J., Gil, Y., Groth, P., Kwasnikowska, N., Miles, S., Missier, P., Myers, J., Plale, B., Simmhan, Y., Stephan, E., Bussche, J.V.D.: The Open Provenance Model core specification (v1.1). Future Generation Computer Systems 27(6), 743–756 (2010) ISSN 0167-739X, 10.1016/j.future.2010.07.005
8. Groth, P., Miles, S., Moreau, L.: PReServ: Provenance Recording for Services. In: UK e-Science All Hands Meeting 2005, Nottingham, UK (September 2005)
9. Groth, P., Moreau, L.: Recording Process Documentation for Provenance. IEEE Trans. Parallel Distrib. Syst. 20(9), 1246–1259 (2009)
10. Wen, L., Wang, J., van der Aalst, W.M.P., Huang, B., Sun, J.: Mining Process Models with Prime Invisible Tasks. Data and Knowledge Engineering 69(10), 999–1021 (2010)
11. Zhang, J., Liu, Q., Xu, K.: Flow Recommender: a workflow recommendation technique for process provenance. In: Proceedings of the Eighth Australasian Data Mining Conference (AusDM 2009), Brisbane, Australia (December 2009)
12. Koop, D., Scheidegger, C.E., Callahan, S.P., Freire, J., Silva, C.T.: Viscomplete: Automating Suggesstions for Visualization Pipelines. IEEE Transactions on Visualisation and Computer Graphics 14(6), 1691–1698 (2008)
13. Antonatos, S., Anagnostakis, K., Markatos, E.: Generating realistic workloads for network intrusion detection systems. In: ACM Workshop on Software and Performance, Redwood Shores, CA, USA (2004)
14. Noble, B.D., Satyanarayanan, M., Nguyen, G.T., Katz, R.H.: Trace-Based Mobile Network Emulation. In: Proceedings of SIG-COMM 1997, Cannes, France, pp. 51–61 (September 1997)
15. Curbera, F., Doganata, Y.N., Martens, A., Mukhi, N., Slominski, A.: Business provenance - a technology to increase tracibility of end-to-end operations. In: OTM Conferences, vol. (1), pp. 100–119 (2008)

16. Davidson, S.B., Freire, J.: Provenance and scientific workflows: challenges and opportunities. In: Proceedings of ACM SIGMOD, pp. 1345–1350 (2008)
17. Doganata, Y., Curbera, F.: Effect of Using Automated Auditing Tools on Detecting Compliance Failures in Unmanaged Processes. In: Dayal, U., Eder, J., Koehler, J., Reijers, H.A. (eds.) BPM 2009. LNCS, vol. 5701, pp. 310–326. Springer, Heidelberg (2009)
18. Bodnarchuk, R.R., Bunt, R.B.: A synthetic workload model for a distributed systems file server. In: Proceedings of the SIGMETRICS Interational Conference on Measurement and Modeling of Computer Systems, pp. 50–59 (1991)
19. Mehra, P., Wah, B.: Synthetic Workload Generation for Load-balancing Experiments. IEEE Parallel and Distributed Technology 3(3), 4–19 (1995)
20. Sreenivasan, K., Kleinman, A.J.: On the construction of a representative synthetic workload. Communications of the ACM, 127–133 (1974)
21. Freire, J., Koop, D., Santos, E., Silva, C.T.: Provenance for Computational Tasks: A Survey. Computing in Science and Engineering 10(3), 11–21 (2008)
22. Missier, P., Sahoo, S.S., Zhao, J., Goble, C., Sheth, A.: *Janus*: From Workflows to Semantic Provenance and Linked Open Data. In: McGuinness, D.L., Michaelis, J.R., Moreau, L. (eds.) IPAW 2010. LNCS, vol. 6378, pp. 129–141. Springer, Heidelberg (2010), doi:10.1007/978-3-642-17819-1-16.
23. Frew, J., Janée, G., Slaughter, P.: Automatic Provenance Collection and Publishing in a Science Data Production Environment—Early Results. In: McGuinness, D.L., Michaelis, J.R., Moreau, L. (eds.) IPAW 2010. LNCS, vol. 6378, pp. 27–33. Springer, Heidelberg (2010)
24. Holland, D., Seltzer, M., Braun, U., Muniswamy-Reddy, K.: PASSing the provenance challenge. Concurrency and Computation: Practice and Experience 20(5), 531–540 (2008)
25. Gu, W., Eisenhauer, G., Schwan, K.: Falcon: On-line Monitoring and Steering of Parallel Programs. Concurrency Practice and Experience 10(9), 699–736 (1998)
26. Newhouse, S., Schopf, J., Richards, A., Atkinson, M.: Study of user priorities for e-infrastructure for e- research (SUPER). In: Proc. of the UK e-Science All Hands Conference (2007)
27. Scheidegger, C., Koop, D., Santos, E., Vo, H., Callahan, S., Freire, J., Silva, C.: Tackling the Provenance Challenge one layer at a time. Concurrency and Computation: Practice and Experience 20(5), 473–483 (2008)
28. Bose, R., Frew, J.: Lineage retrieval for scientific data processing: a survey. ACM Comput. Survey 37(1), 1–28 (2005)
29. Moreau, L.: The foundations for provenance on the web. Foundations and Trends in Web Science 2(2-3), 99–241 (2010)
30. Altintas, I., Barney, O., Jaeger-Frank, E.: Provenance Collection Support in the Kepler Scientific Workflow System. In: Moreau, L., Foster, I. (eds.) IPAW 2006. LNCS, vol. 4145, pp. 118–132. Springer, Heidelberg (2006)
31. Oinn, T., Greenwood, M., Addis, M., Alpdemir, N., Ferris, J., Glover, K., Goble, C., Goderis, A., Hull, D., Marvin, D., Li, P., Lord, P., Pocock, M., Senger, M., Stevens, R., Wipat, A., Wroe, C.: Taverna: lessons in creating a workflow environment for the life sciences. Concurrency and Computation: Practice and Experience 18(10), 1067–1100 (2006)
32. Zhao, J., Wroe, C., Goble, C., Stevens, R., Quan, D., Greenwood, M.: Using Semantic Web Technologies for Representing E-science Provenance. In: McIlraith, S.A., Plexousakis, D., van Harmelen, F. (eds.) ISWC 2004. LNCS, vol. 3298, pp. 92–106. Springer, Heidelberg (2004)

33. Zhao, J., Goble, C., Stevens, R., Turi, D.: Mining Taverna's semantic web of provenance. Concurrency and Computation: Practice and Experience 20(5), 463–472 (2008)
34. Barga, R., Simmhan, Y., Withana, E.C., Sahoo, S., Jackson, J.: Provenance for Scientific Workflows Towards Reproducible Research. Bulletin of Technical Committee on Data Engineering, Special Issue on Data Provenance 33(3), 50–58 (2010)
35. Valerio, M., Sahoo, S., Barga, R., Jackson, J.: Capturing Workflow Event Data for Monitoring, Performance Analysis, and Management of Scientific Workflows. In: IEEE Fourth Int'l Conf. on e-Science 2008 (e-Science 2008), pp. 626–633 (2008)
36. Miles, S., Groth, P., Branco, M., Moreau, L.: The requirements of recording and using provenance in e-science experiments. Journal of Grid Computing 5(1), 1–25 (2007)
37. PC3, `http://twiki.ipaw.info/bin/view/Challenge/ThirdProvenanceChallenge` (accessed December 20, 2009)
38. Data to Insight Center, `http://pti.iu.edu/d2i/provenance-karma` (accessed January 2011)
39. RabbitMQ Messaging System, `http://www.rabbitmq.com` (accessed July 2011)
40. The WS-BPEL Extension for People (BPEL4People), Version 1.0 Specification, `http://www.ibm.com/developerworks/webservices/library/specification/ws-bpel4people` (accessed December 2011)
41. Cao, B., Plale, B., Subramanian, G., Robertson, E., Simmhan, Y.: Provenance Information Model of Karma. In: IEEE Third Int'l Workshop on Scientific Workflows (SWF 2009), Los Angeles, CA (July 2009)
42. Mukhi, N.K.: Monitoring Unmanaged Business Processes. In: Meersman, R., Dillon, T.S., Herrero, P. (eds.) OTM 2010. LNCS, vol. 6426, pp. 44–59. Springer, Heidelberg (2010)
43. Doganata, Y., Curbera, F.: Effect of Using Automated Auditing Tools on Detecting Compliance Failures in Unmanaged Processes. In: Dayal, U., Eder, J., Koehler, J., Reijers, H.A. (eds.) BPM 2009. LNCS, vol. 5701, pp. 310–326. Springer, Heidelberg (2009)
44. Plale, B., Cao, B., Aktas, M.: Provenance Collection of Unmanaged Workflows with Karma. Journal Manuscript Accepted with Revisions (July 2011)
45. Shirasuna, S.: A Dynamic Scientific Workflow System for the Web Services Architecture. PhD thesis, Indiana University (September 2007)
46. Gil, Y., Ratnakar, V., Deelman, E., Mehta, G., Kim, J.: Wings for Pegasus: Creating Large-Scale Scientific Applications Using Semantic Representations of Computational Workflows, pp. 1767–1774. AAAI (2007)
47. Kim, J., Gil, Y., Spraragen, M.: Principles for interactive acquisition and validation of workflows. J. Exp. Theor. Artif. Intell. 22(2), 103–134 (2010)
48. Leake, D.B.: Case-Based Reasoning in Context: The Present and Future. In: Leake, D.B. (ed.) Case-Based Reasoning: Experiences, Lessons, and Future Directions, pp. 1–35. AAAI Press/MIT Press (1996)
49. de Mántaras, R.L., McSherry, D., Bridge, D.G., Leake, D.B., Smyth, B., Craw, S., Faltings, B., Maher, M.L., Cox, M.T., Forbus, K.D., Keane, M.T., Aamodt, A., Watson, I.D.: Retrieval, reuse, revision and retention in case-based reasoning. Knowledge Eng. Review 20(3), 215–240 (2005)
50. Leake, D.B., Kendall-Morwick, J.: Towards Case-Based Support for e-Science Workflow Generation by Mining Provenance. In: Althoff, K.-D., Bergmann, R., Minor, M., Hanft, A. (eds.) ECCBR 2008. LNCS (LNAI), vol. 5239, pp. 269–283. Springer, Heidelberg (2008)
51. Leake, D., Kendall-Morwick, J.: Four Heads Are Better than One: Combining Suggestions for Case Adaptation. In: McGinty, L., Wilson, D.C. (eds.) ICCBR 2009. LNCS, vol. 5650, pp. 165–179. Springer, Heidelberg (2009)

52. Cheah, Y.-W., Plale, B., Kendall-Morwick, J., Leake, D., Ramakrishnan, L.: A Noisy 10GB Provenance Database. In: Second International Workshop on Traceability and Compliance of Semi-Structured Processes, Clermont-Ferrand, France (2011) (in press)
53. Kendall-Morwick, J., Leake, D.: A Toolkit for Representation and Retrieval of Structured Cases. In: Proceedings of the ICCBR 2011 Workshop on Process-Oriented Case-Based Reasoning, Greenwich, U.K. (2011) (in press)
54. Ramakrishnan, L., Plale, B., Gannon, D.: WORKEM: Representing and Emulating Distributed Scientific Workflow Execution State. In: Proceedings of the 10th IEEE/ACM Int'l Symposium on Cluster, Cloud and Grid Computing (CCGrid 2010), Melbourne Australia (2010)

Part II
Data Provenance and Data Management Systems

Chapter 4
Sketching Distributed Data Provenance

Tanu Malik, Ashish Gehani, Dawood Tariq, and Fareed Zaffar

Abstract. Users can determine the precise origins of their data by collecting detailed provenance records. However, auditing at a finer grain produces large amounts of metadata. To efficiently manage the collected provenance, several provenance management systems, including SPADE, record provenance on the hosts where it is generated. Distributed provenance raises the issue of efficient reconstruction during the query phase. Recursively querying provenance metadata or computing its transitive closure is known to have limited scalability and cannot be used for large provenance graphs. We present *matrix filters*, which are novel data structures for representing graph information, and demonstrate their utility for improving query efficiency with experiments on provenance metadata gathered while executing distributed workflow applications.

4.1 Introduction

The provenance of data is a description of how the data came into being or was derived. Provenance metadata is becoming increasingly useful in addressing a wide variety of issues, such as performance optimization, generating repeatable and reproducible scientific computation, security verification, and policy validation for checking regulatory compliance. Consequently, applications are being coupled with

Tanu Malik
University of Chicago, Illinois 60637, USA
e-mail: tanum@ci.uchicago.edu

Ashish Gehani · Dawood Tariq
SRI International, California 94025, USA
e-mail: {ashish.gehani,dawood.tariq}@sri.com

Fareed Zaffar
Lahore University of Management Sciences, Punjab 54792, Pakistan
e-mail: fareed.zaffar@lums.edu.pk

Q. Liu et al. (Eds.): Data Provenance and Data Management in eScience, SCI 426, pp. 85–107.
springerlink.com

suitable provenance middleware that can audit events, read logs, and answer provenance-related questions.

We are particularly interested in provenance infrastructure that is used with applications that perform distributed computation. In this context, consider some examples that give rise to a variety of interesting issues: (i) scientific applications decompose data-intensive problems into subtasks and distribute them across a Grid through a workflow planner that may not track provenance; (ii) scientists who conduct distributed experimental analyses on a variety of research hardware, such as mass spectroscopes, DNA sequencers, or oscilloscopes, must maintain records of the combined analyses for reproducibility; (iii) when different users share data through network connections, the resulting information generated has distributed provenance that may be drawn from multiple, independent administrative domains.

A characteristic feature of such distributed applications is that they are often conducted in loosely controlled environments and use heterogeneous software platforms. It is therefore important to collect such provenance metadata in an application-agnostic manner. The Open Provenance Model (OPM) provides a specification that serves this purpose and allows provenance to be exchanged between systems through a generic vocabulary [27]. Tracking distributed computations at the operating system level allows coupling between the filesystem's state and the associated provenance metadata [32, 11]. A significant implication of this design choice, however, is that it results in large volumes of provenance metadata [12]. Nevertheless, a number of systems, including PASS and SPADE, support transforming such provenance records into OPM.

Provenance systems that audit at fine granularity employ various architectures and mechanisms to manage the resulting metadata. Several systems [32, 4, 36] collect provenance information in centrally managed databases, often referred to as provenance stores. Benefits of aggregating provenance information in central stores include the ease of maintenance and curation, storage efficiency, and access control [17]. These mechanisms, however, also introduce significant network overhead, with many provenance records being transferred to the central provenance store, although remote queries for them may never arise [12]. Accordingly, it is important for distributed applications to account for the location where provenance metadata is collected, processed, stored, and consumed.

Support for Provenance Auditing in Distributed Environments, `SPADE` [37] is a data provenance management system. `SPADE`v2 refers to the second generation of the system, which has modular components for gathering, integrating, filtering, storing, and querying data provenance. Except for the components that gather provenance, the rest are completely agnostic to the source domain. `SPADE` uses Reporter modules customized to the provenance domain to transform the specific semantics into an OPM compliant form. The domain can be a particular application, the operating system, or even manual curation. To manage the resulting provenance, `SPADE` embodies a decentralized model, with each distributed host maintaining the authoritative repository of provenance metadata collected on it. `SPADE`v2's modules for tracking operating system activity record not only data flow dependencies between files and processes but also data movement across systems via network connections. All provenance information is stored in a local database.

Distributed provenance management systems, such as SPADE, face a significant challenge when reconstructing data provenance that spans multiple hosts. The problem is often solved by tracing a path or recursively querying metadata that is manifested as a directed graph. Recursive querying is known to have poor response times for large provenance graphs [20]. In the case of distributed provenance, it is also expensive in terms of network operations since the provenance metadata is unlikely to be located where the data is stored, and the appropriate remote sources must be identified. The alternative to recursive querying is computing a transitive closure, which is computationally expensive. In addition, this requires global knowledge, which raises traditional distributed system challenges.

SPADE employs *provenance sketches* to address the problem of reconstructing distributed data provenance. Such provenance can be viewed as a collection of subgraphs, each from a different host, that interface through vertices corresponding to network connections between the hosts. The provenance sketches determine which network connections are relevant to a query, while locally computed transitive closures provide host-specific subgraphs that must then be stitched together. In our earlier work [24], provenance sketches summarized host-specific provenance subgraphs with Bloom filters [2]. In contrast, we now encode an entire provenance graph by organizing a set of Bloom filters into a new data structure that we term a *matrix filter*. Matrix filters, when propagated to other downstream hosts, determine in a single lookup the existence of a path between any two distributed hosts, which would previously have required contacting multiple hosts. If the path exists, the matrix filter can also be used to determine the specific remote hosts that contain the intermediate path. This allows us to contact the intermediate remote hosts in parallel to construct the full provenance path rather than building the path one remote host at a time. The parallel operation substantially improves the performance of distributed path queries.

We deployed SPADE to collect fine-grained provenance of workflows used in the NIGHTINGALE project [30]. The project uses heterogeneous machine learning algorithms to translate information from multiple languages so that monolingual users can query the content. The provenance of intermediate outputs is used when comparing the quality of competing approaches. We mapped the provenance metadata to distributed SPADE databases, and constructed representative provenance queries. SPADE was augmented with functionality to compute the provenance sketches needed for each host. Our experiments indicate that queries are answered accurately with the aid of matrix filters. Query response times remain constant even when the number of levels in the provenance increases.

The remainder of the chapter is organized as follows. Section 4.2 describes provenance systems for distributed applications. Section 4.3 outlines the SPADE architecture and data model for auditing system-level provenance and storing it in distributed repositories. Section 4.4 describes sketches for encoding graphs. In particular, it describes the matrix filter and how it can be used for improving the latency of provenance queries in a distributed provenance system, such as SPADE. Section

4.5 reports our findings about the use of matrix filters to improve the efficiency of SPADE queries in a PlanetLab [31] distributed environment. Section 4.6 concludes.

4.2 Related Work

Distributed applications manage and query digital provenance in a variety of ways. Chimera [8] uses a virtual data catalog to store information about Grid data objects, transformation types, and applications. Swift extends concepts in Chimera to include a custom provenance data store with an SQL-like language [10]. ES3 [9, 25] and PASOA [14, 16] record the provenance of files in distributed services, but provide minimal query interfaces. Karma explores a service-oriented architecture for collecting provenance metadata about workflows. It employs basic recursive traversal to enhance query capabilities [36]. Service-oriented Grids also gather provenance at multiple locations using distributed protocols [15].

One primary issue that arises with distributed data artifacts is how they should be semantically described and referenced. OPM facilitates interoperability between systems by providing a common model for provenance. Several projects provide OPM-compliant provenance, such as SPADE, PASS [33], VisTrails [4], and Tupelo [40]. More recently, an OPM profile (which is a set of conventions) models aspects such as transactions in distributed systems [18].

Not all systems, however, provide a combined comprehensive recording and querying infrastructure. The PASS project developed the provenance query language (PQL) [21]. PQL, however, does not interact with the distributed provenance gathering system PA-NFS [33] that enhances NFS to record provenance in local area networks. ExSPAN [41] allows the exploration of provenance in networked systems. Both systems use provenance metadata to answer queries about the origin of data and how it was derived. The ExSPAN scheme extends traditional relational models for storing and querying provenance metadata, while SPADE supports both graph and relational database storage and querying. Queries in ExSPAN are not optimized for performance.

ProQL [22] is a query language for provenance graphs and presents a convenient way of exploring tuples and nodes, and the ability to isolate and request portions of the graph. Similarly, D-PQuery [17] allows fetching of portions of a provenance graph in a distributed setting. However, the efficiency of queries is not addressed. ExSPAN explores storage and query optimization techniques to reduce communication latency and bandwidth, and employs caching of provenance metadata to improve query performance [41]. Caching assumes locality over the incoming query pattern. SPADE employs summary data structures to improve the performance of each distributed query.

An issue related to distributed provenance querying is the identification of objects uniquely across different administrative boundaries. PASS describes global naming, indexing, and querying in the context of sensor data [34]. SPADE addresses the issue by using storage identifiers for provenance vertices that are unique to a host and requiring distributed provenance queries to disambiguate vertices by referring to them by the host on which the vertex was generated as well as the identifier local to that host.

Given the data-intensive nature of managing provenance metadata, providing adequate storage can be a challenge. PASS explores storing provenance in highly available fault tolerant environments, such as clouds [34, 35]. SPADE employs flexible provenance storage, including graph databases, installed on the hosts where the provenance is generated. Provbase [1] uses Hbase, an open-source implementation of Google's BigTable [5] to store and query scientific workflow provenance. Provenance metadata is exported to Hbase as RDF triples, and SPARQL is used to query Hbase using its native API. Storing provenance concisely has also been investigated elsewhere [41, 20] and remains an active area of research.

4.3 Tracking System-Level Provenance with SPADE

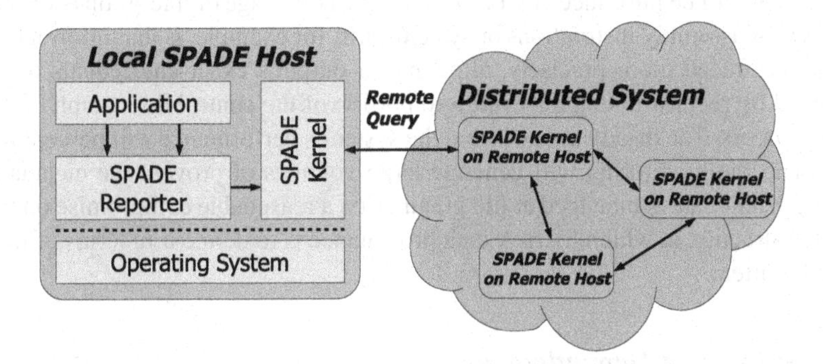

Fig. 4.1 The SPADE kernel provides an independent provenance middleware service on each host in the distributed system. A *SPADE Reporter* is a module that transforms records from a domain, such as operating system activity, into an OPM-compliant representation of data provenance. Distributed provenance queries are transparently handled by the local service, which contacts remote daemons as needed.

SPADEv1 refers to an initial implementation that enabled provenance questions about executed processes and files that were read and written by them. SPADEv2 is a second generation of the system and adopts the OPM model. It has a kernel, storage, querying, and filtering that is agnostic to the source of provenance, with domain-specific annotations created in *Reporter* modules, as illustrated in Figure 4.1. The rest of this chapter focuses on the use of operating system Reporters that audit filesystem reads and writes, process execution, and TCP connections, and transform them to OPM.

In a distributed environment each computer has the freedom to maintain an independent filesystem and accompanying namespace, and yet data can be shared across organizational boundaries. The provenance recording infrastructure must overlay a coherent framework that facilitates reasoning about the origins of data in such a distributed environment. In particular, the infrastructure must track data flows within a host — that is, intra-host dependencies, and across hosts — that is, inter-host dependencies. We now describe how we record both intra-host and inter-host operating system dependencies with `SPADE`.

Recording provenance by tracking data flows requires the system to (i) identify the producers and consumers of each piece of data, and (ii) define the granularity at which a piece of data will be tracked. On a single host, the immediate source of a piece of data will be a process, which may in turn (recursively) have used data written by other processes that have executed on the same host. In addition to the data flowing within a single host, processes may have read data from other hosts through network connections. In such an event, the provenance of any data modified by a process must also include the provenance of the data read from the remote host. We adopt the convention of identifying data by both its location in the system and the time at which it was last modified.

The granularity at which we track the provenance of a data object affects the overhead that will be introduced in the system. The advantage of fine-grain auditing, at the level of assembly instructions or system calls, for example, is that information flow can be traced more precisely, allowing an output's exact antecedents to be ascertained by reconstructing the exercised portion of the control flow graph of the relevant process. The disadvantage is that the system's performance will perceptibly degrade and the monitoring will generate large volumes of provenance metadata. Since persistent data is managed at file granularity, a reasonable compromise on the level of abstraction at which to track data provenance is to define it in terms of files read and written.

4.3.1 Intra-host Dependencies

We utilize the following elements to model intra-host dependencies in a provenance graph:

- *Process vertices* are initialized when the auditing system first encounters a process. Each vertex contains a range of attributes, including the name of the process, its operating system identifier, owner, and group. Each vertex also records the parent process, the host on which the process is running, the creation time of the process, the command line with which it was invoked, and the values of environment variables. We do not version process vertices as the state changes (when an environment variable is updated, for example), although this could be useful for long-running processes such as server daemons.

- *File vertices* include various attributes associated with a file, including the host on which it resides, its pathname in the host's filesystem, the size of the file, the last time it was modified, and optionally a hash of the file's contents and a digital signature by the file's owner to attest the integrity of the hash. When the provenance of a file is being discussed, the *sink* of the associated provenance graph will be the vertex corresponding to the file. We adopt the convention of identifying a file using both its logical location and its last time of modification to disambiguate different versions of the same file, which avoids data dependency cycles in the provenance graph.

- *Edges* in a provenance graph are directed, signifying the direction of the data dependency. An edge to a file vertex indicates that the file was read, while an edge from a file vertex indicates that the file had been modified. Analogously, an edge from a process indicates that a read operation was performed by the process, while an edge to a process vertex reflects a write operation. Consequently, read and write operations to and from the filesystem by a process can be modeled by a provenance graph.

In the context of provenance, we define the semantics of a *primitive operation* to be an output file, the process that generated it, and the set of input files it read in the course of its execution. For example, if a program reads a number of data sets from disk, computes a result and records it in a file, a primitive operation has been performed. If a process modifies a number of files, a separate instance of the representation is used for each output file.

Primitive operations are combined into a *compound operation*. For instance, if the result of appending together several data sets (by a program such as UNIX *cat*) is then sorted into a particular order (using another program, such as UNIX *sort*, that executes as a separate process), then the combination of appending and sorting is a compound operation. Thus, the provenance of every file can be represented by a compound operation that is a directed graph, consistent with the model used by Grid projects [39].

4.3.2 Inter-host Dependencies

We now consider a simple example where an operation spans multiple hosts. A user with identity **501** on the machine with IP address **10.12.0.55** uses *ssh* to connect to a remote host. The user runs the UNIX *cat* program to output the contents of the file **/var/log/remote_httpd.log**. The output is redirected into the file **/tmp/local_httpd.log** in the filesystem of the host where the *ssh* command was invoked. This effectively copies the contents of the remote file to the local file.

```
% ssh 501@10.12.0.55 cat /var/log/remote_httpd.log > /tmp/local_httpd.log
```

Similar commands and analogous file transfer utilities like *sftp*, FTP, or GridFTP are commonly used in large distributed computations to move input data to idle processors and to retrieve the results after the execution completes. If the provenance tracking was restricted to inter-host dependencies, queries about the provenance of

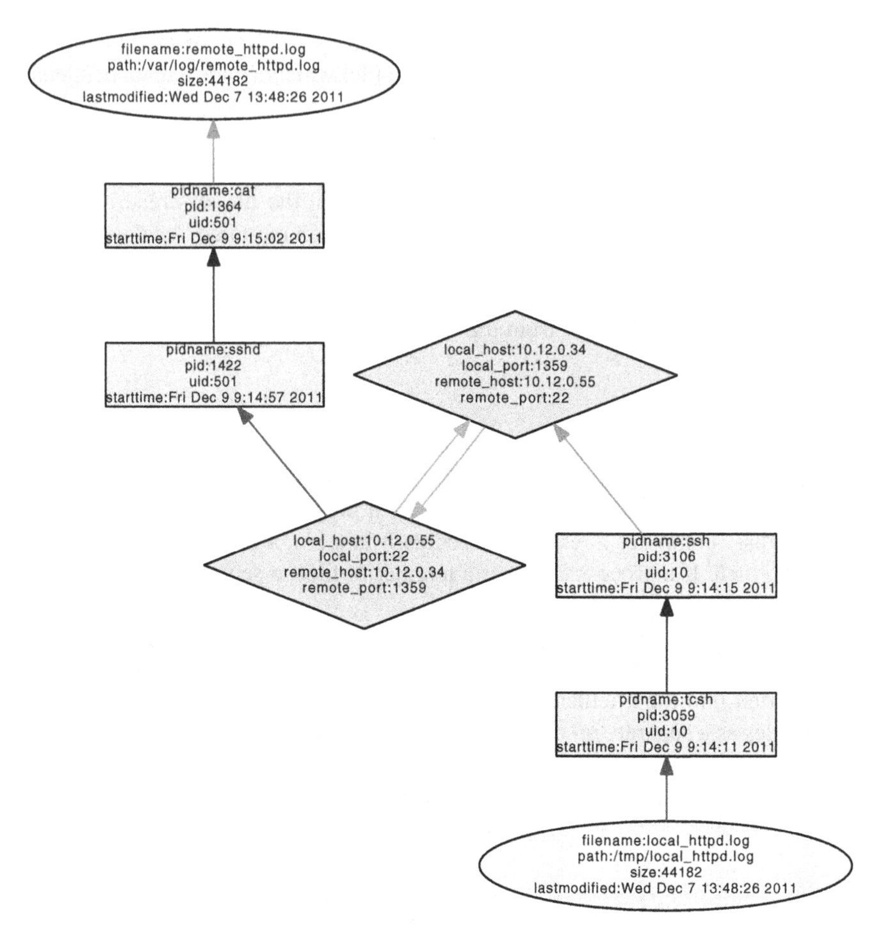

Fig. 4.2 A vertex shown with a rectangle represents the execution of a process, while a vertex shown with an ellipse represents a file that was read or written. A network vertex, depicted using a diamond, has the property that its attributes can *independently* be inferred at both ends of a connection. TCP connections used for protocols such as ssh, FTP, HTTP, or Java RMI allow the construction of such network vertices.

the file **/tmp/local_httpd.log** would not be able to establish a relationship to the file **/var/log/remote_httpd.log** on the machine **10.12.0.55**.

One approach to addressing the gap described above is to record information about the host on which each process runs and where each file is located. Users can then be provided a mechanism for transferring the provenance metadata when a file moves from one computer to another. Records that refer to the part of the provenance graph that originated on a remote host will be explicitly disambiguated using the *host* attribute. While this scheme ensures that all provenance queries can be answered at the destination host, it incurs considerable storage overhead [11].

An alternate approach would avoid replicating the provenance records at the destination host to which the file is being transferred. Instead, the provenance store at the destination would be provided with a pointer back to the relevant provenance metadata on the source host. However, provenance queries at the destination would require the source hosts to be contacted, slowing the response time and decreasing reliability (since remote hosts may be unreachable).

In the above example, a distributed data flow takes the form of a file transfer. In practice, data may also flow through network connections directly from one process to another, as is the case in service-oriented architectures. In such systems, a series of HTTP calls is made from one host to another, each passing XML documents that include requests and arguments, and corresponding XML responses with return values.

- To model network flows, we introduce a fourth type of element in provenance graphs — the *network vertex* — that has the property that its attributes can *independently* be inferred in two or more processes. If the processes are being audited by different provenance middleware, the property ensures that each system can construct an equivalent network vertex without any explicit coordination. For example, equivalent network vertices associated with a TCP connection can be constructed at both endpoints using the local IP address and TCP port, remote IP address and TCP port, and timestamp (including the date), as illustrated in Figure 4.2.

Figure 4.2 depicts the provenance graph for the file **/tmp/local_httpd.log** that would arise after execution of the *ssh* command described earlier. (The graph is simplified for clarity.) The key point to note is that the provenance vertex for the network connection (between *ssh* and *sshd* in the example) can be independently constructed by both the hosts at the two ends of the network connection. This allows complete decentralization of the provenance recording in the distributed system, with each host's provenance infrastructure operating independently. At the same time, the provenance records generated can be pieced together to yield a coherent and complete reconstruction of the distributed data flows.

4.4 Querying Provenance

SPADE provides a query client that can be used to inspect the provenance metadata generated by the operating-system-level Reporters to ask questions about processes that ran, the files they read or wrote, and the network connections they initiated or handled. In particular, this can be used to answer the questions asked in the First Provenance Challenge of the International Provenance and Annotation Workshop (IPAW) [26]. SPADEv2 supports a variety of storage formats for the provenance metadata. This includes the default storage in the graph database Neo4j [29], the embedded relational database H2 [19] (and with minor changes, MySQL [28]). The client can interrogate each storage with the underlying database's query language, as well as custom provenance queries.

All of the domain-specific semantics are recorded as annotations of the OPM vertices. SPADEv2 requires the user to specify the hosts on which the known elements of the query are present. It maps these elements to globally unique provenance identifiers in an initial phase, and then uses its auxiliary data structures to operate on provenance graphs that are represented in terms of these identifiers.

The provenance queries in the IPAW challenge can be classified into those that require access to (i) the entire provenance graph of the output file, (ii) just a subgraph of the provenance of a vertex, or (iii) a path in the provenance graph between specific input and output vertices. These categories lead to the following query specifications: (a) given a vertex, request its entire provenance graph, (b) path expressions with vertex attributes that include process and file identifiers, (c) given a vertex, request its provenance subgraph up to k levels, and (d) check if a path exists between two vertices, s and t. We will focus on provenance path queries of type (d) as they are the most general and often have high latency.

Provenance path queries can be answered recursively – by following a pointer, corresponding to the direction from which data had flowed – or by computing the transitive closure over the entire graph. It has been shown experimentally that standard recursive graph traversal algorithms do not scale for large workflow processes and for large collections of data sets [20]. The alternative method of computing the transitive closure over the entire provenance graph is computationally expensive and has a large storage overhead [7]. When the graph is distributed, computing the transitive closure is a complex operation. An efficient method for computing the transitive closure has been described [20], but it is not clear how it translates to a decentralized scenario.

We adopt a hybrid approach for answering distributed provenance path queries. Across distributed hosts, the query is computed recursively. Within a host, the query is computed using the transitive closure, an operation that is natively supported if the storage used is a graph database.

We improve the efficiency of recursive querying with sketches that help reduce the number of hosts that must be contacted when constructing the response to a provenance path query. The sketches are space-efficient representations of graphs and are used in SPADE to track connections between network vertices. In the rest of this section, we describe how provenance sketches are constructed and how they are used in SPADE to efficiently answer provenance path queries.

4.4.1 Provenance Sketches

Consider a graph, such as the one at the top of Figure 4.3, that depicts the provenance of a piece of data on a single host in terms of identifiers for file and process vertices and edges representing the data dependencies. We introduced the notion of a provenance sketch [13] to allow such a graph to be succinctly represented. Depending on how the sketch is constructed, it can support a specific set of queries.

The sketches we will describe use Bloom filters as building blocks. A Bloom filter is a compact data structure that provides a probabilistic representation of a set.

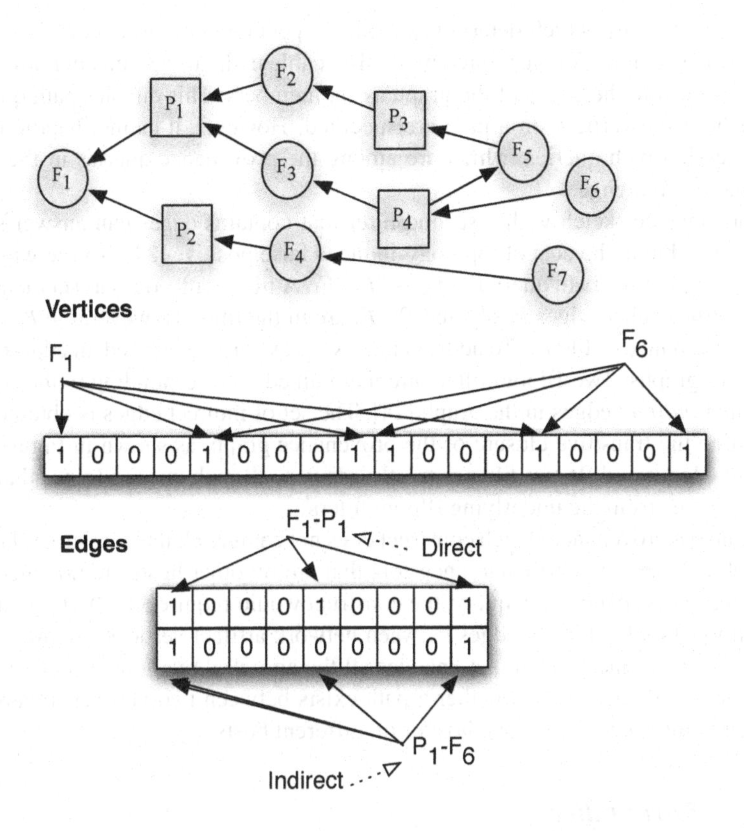

Fig. 4.3 A sketch of a provenance graph can created by inserting the hash of each vertex in a Bloom filter, the hash of each direct edge in a second Bloom filter, and the hash of each indrect edge in a third Bloom filter. However, this construction has limitations.

It supports membership queries – that is, queries that ask "Is element X in set Y?", denoted with the predicate $inSet(Y,X)$. Given a set $A = a_1,\ldots,a_n$ of n elements, a Bloom filter uses a vector v of m bits, with all bits initially set to 0, and k independent hash functions, $h_1,\ldots,h_k$, each with a range $1,\ldots,m$. For each element $a \in A$, the bits at positions $h_1(a),\ldots,h_k(a)$ in v are set to 1. Inserting an element only requires it to be hashed with the k functions. Each of the k outputs determines a bit in the vector v that should be set to 1 (if it is currently 0). Given a query for b, the bits at positions $h_1(b),\ldots,h_k(b)$ are checked. If any of them is 0, then certainly b is not in the set S. Otherwise, b is conjectured to be in the set with probability $(1-e^{-kn/m})^k$. The possibility of a false positive is what causes this probability to differ from 1.

We considered a number of methods to encode graphs with Bloom filters, each enabling a different set of queries. The simplest approach is to use the vertices of the graph as elements of the set S [13]. However, this only enables set membership queries, such as $inSet(S,F_1)$ and $inSet(S,P_1)$ where F_1 is a file vertex and P_1 is a process vertex. This cannot support queries about a path in a provenance graph, such as $P_1 / F_2 / \star / P_2 / F_4$, which determines if file F_2 generated by process P_1 is part of the provenance of file F_4, generated by P_2. To enable path queries, an alternative approach is to store the edges of the graph as set members. This enables path queries in which all the vertices on a path are specified. However, it cannot handle regular expression path queries, which are among the provenance queries in the First Provenance Challenge.

A provenance sketch with a second filter that contains edges can answer some path queries but at the cost of topology-induced false positive [24]. To see why this occurs, consider the path query $P_1 / F_2 / \star / P_2 / F_4$. A filter with edges as set members returns "true", since edges $P_1 - F_2$ and $P_2 - F_4$ are in the filter. However, file F_2 is not in the provenance of file F_4. To address this, we previously proposed an edge-based sketch for graphs. Two Bloom filters are maintained – corresponding to the sets of *direct* and *indirect* edges in the graph [24]. The set of indirect edges is obtained by computing the transitive closure of the provenance graph, as shown in Figure 4.3. Such an edge-based Bloom filter correctly answers all path queries (other than the false positives from the underlying Bloom filters).

The above provenance sketch construct does not capture all the ancestral relationships of a vertex – in particular, ancestors that are on other hosts are not encoded. To answer a distributed path query, the construct would require `SPADE` to maintain an additional table of cross-edges between network artifact vertices. Below we introduce a provenance sketch that encodes all the ancestral relationships of vertices and supports queries about whether a path exists between two vertices, regardless of whether they are on the same host or on different hosts.

4.4.2 Matrix Filter

We introduce the *matrix filter*, a new data structure to probabilistically represent graph connectivity (or any other data that can be stored in a matrix). Whereas the original matrix may have a size of $\mu \times \mu$ for an arbitrary μ, the filter only uses $O(m \times m)$ space for a fixed m, thereby providing a compact representation of the matrix.

A matrix filter consists of (a) a *row array* of m bits, (b) a *column array* of m^2 bits, and (c) k independent hash functions $\{h_1, \ldots, h_k\}$, each of which has a range of $\{1, \ldots, m\}$. Further, the ith bit b_i of the row array defined in (a) is associated with the ith set of m bits in the column array defined in (b).

Assume that $\mathscr{S}$ is the set of direct edges in the provenance graph $G = (V, E)$. During an initial setup phase, the edges in $\mathscr{S}$ are inserted into the filter. Consider the insertion of a direct edge $(s,t) \in E$ – that is, s is the parent of t – into the matrix filter, as showin in Figure 4.4(b). First k hash functions are applied to t and each resulting value sets the bits in the row array defined above in (a). Thus, $H_k(t)$ sets one or more of the m bits. For each bit in the row array that is set by $H_k(t)$, k hash functions are applied to s. Each resulting value sets one or more of the ith set of m bits in the column array. Figure 4.4(c) shows that when edge (s,t) is inserted into the empty filter, $H_k(t)$ sets the bits in positions 1, 5, 7. Each of these bits is

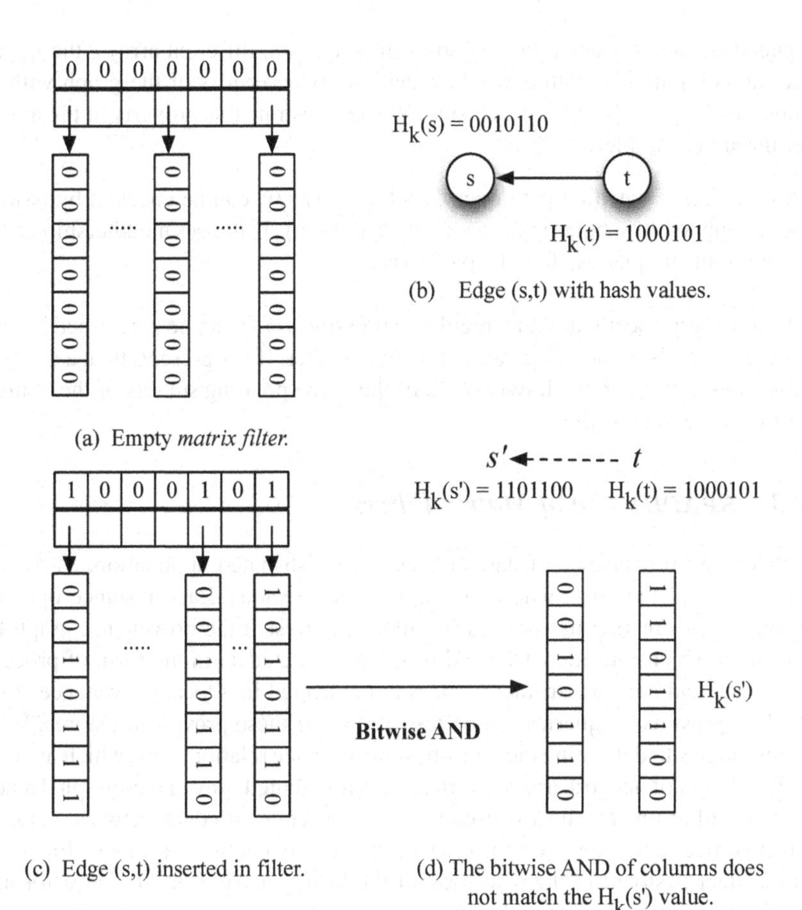

Fig. 4.4 A graph edge (s,t) is stored in a *matrix filter*. The filter is then used to determine that edge (s',t) is not in the graph.

associated with a set of m bits, which is updated with the value of the hash of s, $H_k(s) = 0010110$.

To check if an edge, (s',t), is present in the graph, the k hash functions are applied to t. The bits indexed by $H_k(t)$ are checked in the row array. If they are all set to 1, t is potentially in the graph. Each such bit is associated with a set of m bits in the column array. All these sets of m bits are combined by a bitwise AND. A vertex s' is potentially an ancestor of t if all the bits determined by the hashes $H_k(s')$ are in the bitwise AND. If any of the row array bits is set to 0, this indicates that t is not a child vertex. If any of the bits in the computed bitwise AND is 0 but was pointed to by one of the hashes of s', this indicates that s' is not a parent vertex of t. In Figure 4.4(d), the test for edge (s',t) fails and thus the edge is not in the graph.

In the matrix filter, vertices of an edge are hashed to distinct arrays – one vertex to the row array and the other to the column arrays indexed by the set row array bits.

By encoding the ancestry relationship of the edge into different arrays, the matrix filter can compute if a path exists between any two vertices of the graph without computing the graph's transitive closure. We demonstrate this property of the matrix filter through examples:

- A completely specified path query, such as $s/t/u/v$, can be checked by issuing set membership subqueries (s,t), (t,u), and (u,v). If the set membership test is true for all subqueries, then the path exists.

- A path query specified with a regular expression $s/*/t$ can be performed by first checking if t is in the row array of the matrix filter. If t is present, then a check is done to see if s is in the bitwise AND of the corresponding subsets of the column array of the matrix filter.

4.4.3 SPADE's Use of Matrix Filters

To collect the provenance of data created by a distributed application, SPADE is deployed on all the hosts where the application executes. Each instance operates independently, creating its own matrix filter to represent the provenance graph for the host on which it resides. Detailed provenance metadata in the form of process, file artifact, and network artifact vertices are collected in a local provenance store. Intra-host provenance queries can be resolved using these provenance stores. What remains then is to determine the cross-host provenance relationships, which are captured by the set of network artifact vertices in a distributed provenance graph. Hence, the matrix filter on each host is used to store the set $\mathscr{S}$ of edges between network artifact vertices and their ancestor vertices that are also network artifacts. Each network artifact is stored in the row array of the host's matrix filter. Its ancestor network artifact vertices are stored in the corresponding column array locations. This includes ancestors on the same host as well as on other hosts. Figure 4.5 shows the sets $\mathscr{S}$ of edges between network artifacts that the hosts insert in their respective matrix filters.

Provenance path queries whose end points reside on different hosts must determine the exact hosts through which their path traverses. Otherwise, computing them will result in a commensurate number of (high latency) network connections on several hosts. Since network artifacts connect hosts, we observe that storing edges only between network artifacts is sufficient to answer distributed provenance path queries. We elucidate this observation with a concrete example in the next subsection.

When a network artifact is generated on a host, SPADE adds it to the provenance store. If the new vertex is associated with an incoming network connection, the SPADE server on the remote host is contacted and that host's matrix filter is retrieved. The set of ancestor network vertices of the new vertex is extracted from the remote host's matrix filter. The new vertex and the set of ancestor vertices are added to the local matrix filter. (A performance optimization was also implemented where remote matrix filters are locally cached to avoid repeated retrieval at the cost

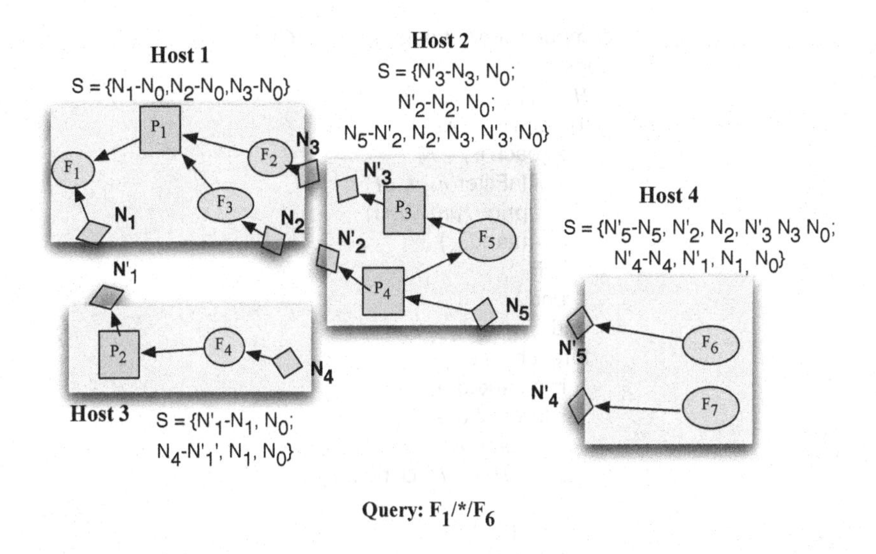

Fig. 4.5 A provenance graph can be distributed across multiple hosts. Query $F_1/*/F_6$ arrives at host 4. The list of all hosts that must be contacted is completely determined locally using host 4's matrix filter.

of losing distributed consistency.) Consequently, the matrix filter of a host includes the ancestral network vertices from all upstream hosts.

A matrix filter construction also requires a judicious choice of the value of k and m. In general, a smaller k is preferred since it reduces the amount of computation [2]. The appropriate choice of m depends on the number of network connections being made. For applications with limited network connectivity, a smaller m in the range 10-50 provides low falses. Similarly, a higher m in the range of 100-500 provides low false positives for more network intensive distributed applications.

4.4.4 Querying Provenance across Multiple Hosts

To illustrate how SPADE handles queries about provenance metadata that spans multiple hosts, we consider the case when the path is a regular expression that includes vertices s and t from different hosts. SPADE tackles distributed provenance path queries in two stages.

In the first step, the SPADE daemon on the host where the query arrives inspects its local matrix filter. The daemon determines if a path exists between the network artifacts that are the descendants and ancestors of vertices s and t, respectively. If such a path exists, SPADE determines the list of potential intermediate hosts that the path traverses. In the second step, the SPADE daemon on each of the hosts in the list is contacted to retrieve a part of the path that satisfies the query specification.

```
compute_query_hosts(Ns, Nt, M, C)
begin
  H ← {}
  foreach n_s ∈ Ns
    foreach n_t ∈ Nt
      if inFilter(n_s, n_t, M)
        print(Path exists)
        mark(n_s)
      fi
    end
  end
  foreach n_s ∈ Ns
    if isMarked(n_s)
      foreach c ∈ C
        if n_s ∈ c
          H ← H U host(c)
        fi
      end
    fi
  end
end
```

Fig. 4.6 M is the matrix filter and C is the cache of sketches, both at the host where t is located. H is the set of hosts that will be contacted for path fragments. In the first stage, every network vertex n_s is marked if there is a descendant network vertex n_t in the matrix filter. In the second stage, a list is built of the hosts with sketches that contain any of the marked network vertices.

Assuming the existence of a provenance graph with components distributed across multiple hosts, each of which has its own matrix filter, we can perform a distributed provenance path query $s/*/t$ as follows:

1. Query the host where s is located to determine N_s, the set of network artifacts that are the descendants of s, and N_t, the set of network artifacts that are ancestors of t.

2. Execute the algorithm in Figure 4.6 to determine the list of hosts that need to be contacted.

If a path exists between the two distributed vertices, the provenance subgraphs corresponding to the query specification must be obtained. To determine the subgraphs, queries are sent to relevant remote hosts in parallel. The daemon that initiated the query receives the path fragments in response and assembles them into a single path from s to t. False positives from a sketch may result in extra path fragments in the response, necessitating careful selection of the parameter m when initializing the matrix filters.

Figure 4.5 illustrates the process of determining which hosts need to be contacted to obtain path fragments in response to a query. In this example, the query $F_1/*/F_6$ arrives at *Host 4*. The daemon on *Host 4* queries the daemon *Host 1* to obtain $N_{F_1} = \{N_1, N_2, N_3\}$, the network artifacts that are the descendants of F_1, and locally determines $N_{F_6} = \{N'_5\}$, the network artifacts that are ancestors of F_6. Using the local sketch, a check is performed to see if any path in the set $\{N_1/N'_5, N_2/N'_5, N_3/N'_5\}$ exists. The paths $\{N_2/N'_5, N_3/N'_5\}$ are present. The cached sketches are used to decide which hosts have network vertices in N_{F_1} as ancestors. If the sketch produces false positives, a small number of unnecessary network connections may still be made.

4.5 Experimental Results

Our experiments are conducted on provenance metadata that was gathered by using SPADE to monitor the workflow of a large distributed application in SRI International's Speech Technology and Research Laboratory [38]. The application workload originated as part of the NIGHTINGALE project [30], which allows monolingual users to query information from newscasts and documents in multiple languages. The objective of the NIGHTINGALE project is to produce an accurate translation. NIGHTINGALE aims to achieve this with a workflow that specifies the tools that will transform the inputs using automatic speech recognition algorithms, machine translation between languages, and distillation to extract responses to queries. However, there is no canonical algorithm for each of these steps, necessitating a choice between a variety of tools. The speech scientists use accompanying metadata to estimate which combination of available tools will produce the most accurate result. This is further complicated by the fact that the tools have multiple versions and are developed in parallel by experts from 15 universities and corporations. Finally, the choice of which specific version of a tool to use depends on the outcome of previous workflow runs.

A representative application workload executed for roughly half an hour with SPADE collecting provenance metadata about the processes that ran and files that were accessed and modified. The resulting provenance graph had 5256 file vertices, 5948 process vertices, 35948 edges, and a depth of 24 levels. Since the workflow was obtained from a single site (at SRI), we divided it to correspond to a distributed execution over eight geographically diverse hosts with matching network connection entries. These were PlanetLab [31] hosts located at SRI, University of Washington, and Princeton University.

To divide the workload into subgraphs corresponding to a distributed workflow, we used hMETIS [23], a graph partitioning tool. When running hMETIS, we used a high *UBfactor* (in the range of 40 to 50), which specifies that a large imbalance is allowed between partitions during recursive bisection. Since hMETIS partitions the workload by recursively creating bisections, the resulting topology over the distributed system has a tree structure, similar to what would result from a workflow planner. This also results in fewer edges between partitions, consistent with the goal of a distributed workflow planner, such as Pegasus [6]. The longest network path length in this tree was 4.

The resulting partitioned provenance graphs were then deployed on each of the eight PlanetLab hosts as SPADE databases. In addition, we constructed a synthetic workload that consists of a single linear path through nine PlanetLab hosts. The second workload provides a control to understand the effect of the sketches on the network latency of provenance queries, independent of the graph characteristics of the workload (since it consists of a series of network connections sequenced through all the hosts). In both workloads, there are 16 network vertices.

We use matrix filters with 20 bit row filters, 20 bit column filters, and 4 hash functions each. The matrix filters are built by inserting all network vertices created on a host, as well as ancestor network vertices from matrix filters of upstream hosts.

A client requests the provenance through a query, which may originate at any host in the system. A query is considered to be local to a host if the end vertex is on that host. Otherwise, the query is transported to the host where it is local. Matrix filters are used to both determine if a path exists as well as to locate the hosts that may contain parts of the provenance related to a query.

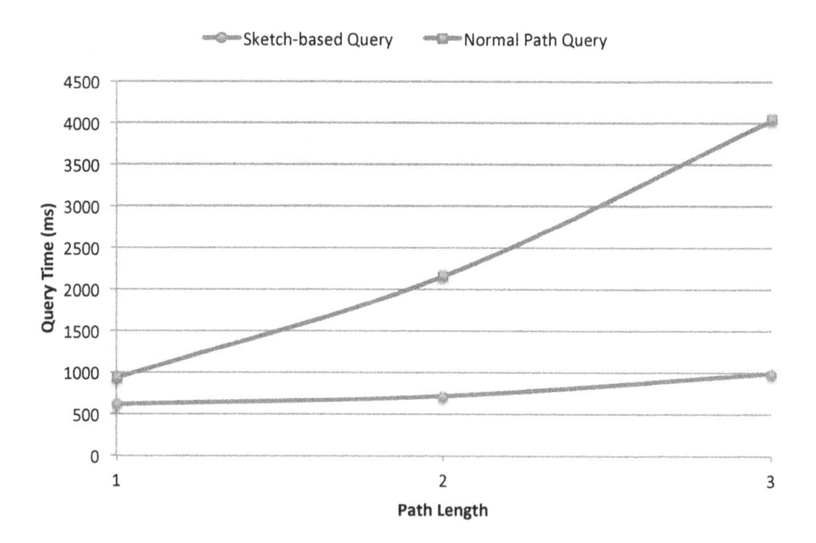

Fig. 4.7 The latency of path queries with and without the use of provenance sketches as a function of the number of hosts that must be contacted. The provenance is of a speech processing workflow.

4.5.1 *Reduction in Network Latency*

The dominant cost of answering distributed provenance path queries comes from the network connections. To measure the reduction in network latency, we undertook the following experiment. The x axis of Figure 4.7 shows the actual number of

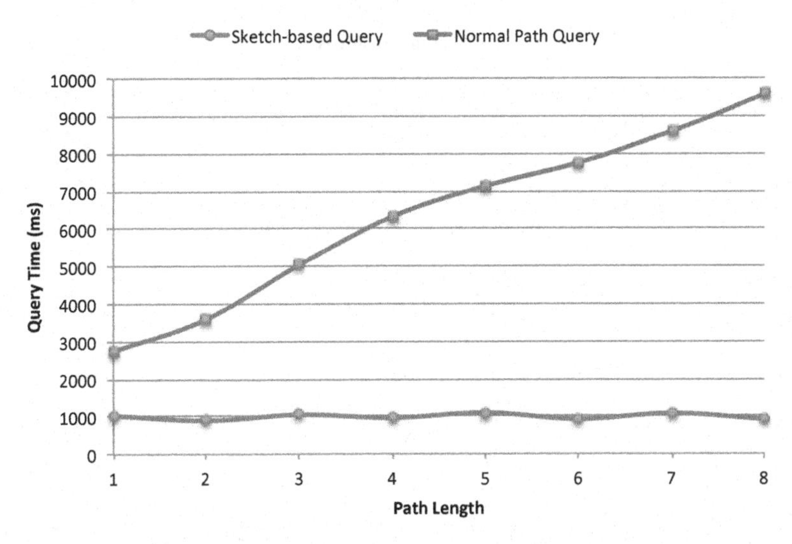

Fig. 4.8 The latency of path queries with and without the use of provenance sketches as a function of the number of hosts that must be contacted. The provenance is of a synthetic workload, consisting of a sequence of network connections through consecutive hosts.

hosts that need to be contacted to respond to a query asking for all the paths from one distributed vertex to another. The time taken to complete the request with and without the use of sketches is shown along the y axis. In particular, note that when using sketches, the latency does not increase when an increasing number of hosts must be contacted. This is true because the sketches allow all the remote hosts to be contacted in parallel, each with a suitable query, corresponding to the piece of the distributed path from that host.

While Figure 4.7 shows that using provenance sketches significantly reduces the time to answer a path query, this is seen even more dramatically in Figure 4.8 with the synthetic workload that demonstrates the effect as the query is scaled to a depth of nine distributed hosts. The latency for answering queries remains constant regardless of the number of remote hosts that must be contacted when sketches are utilized.

4.5.2 Sketch Robustness

To understand the efficacy of our sketches as the provenance graph grows in size, we measured the number of false positive answers to queries about whether an edge between two network vertices exists in a matrix filter. Since the sketch contains provenance metadata of an increasing number of hosts as it propagates along a distributed path in the system, it is expected to provide an increasing number of false positive responses. Figure 4.9 shows that the rate of such false positives is very low, ensuring that the sketches are robust as the provenance grows. Figure 4.10 shows that this is not an issue even in the case of the more strenuous synthetic workload with longer path lengths.

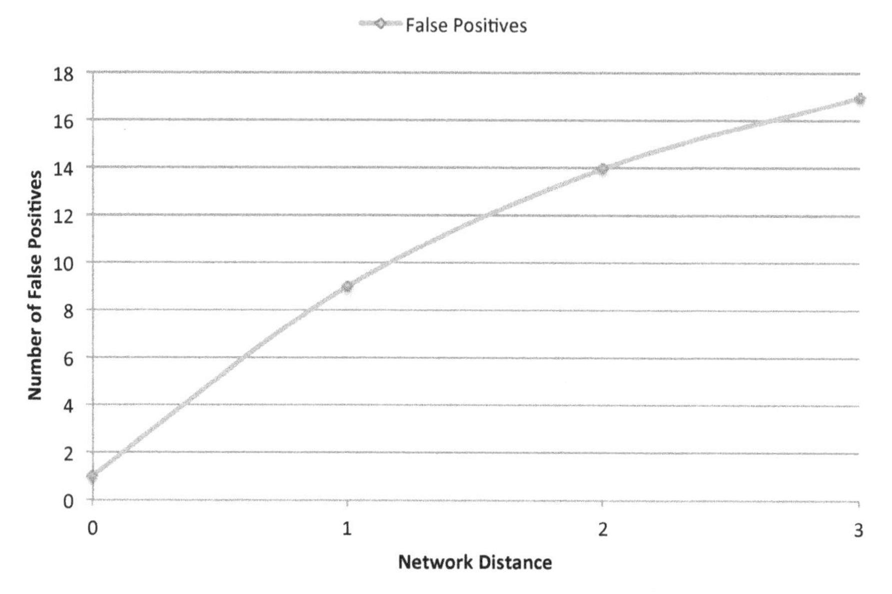

Fig. 4.9 The number of false positive responses in 100,000 random provenance queries. Each query checks if an edge exists between two network vertices using the matrix filter sketch of a host. The provenance is of a speech processing workflow.

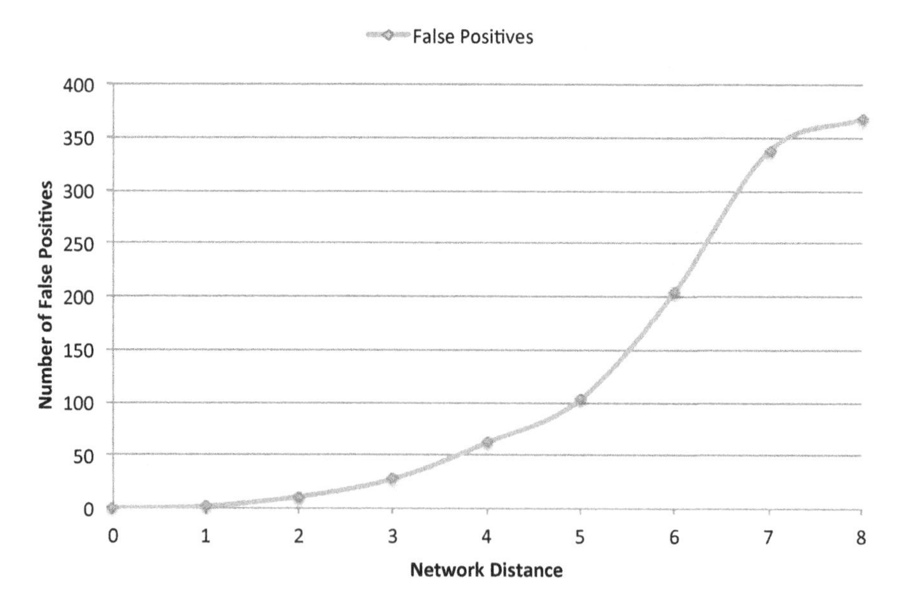

Fig. 4.10 The number of false positive responses in 100,000 random provenance queries checking if an edge exists between two network vertices using the matrix filter sketch of a host. The provenance is of a synthetic workload, consisting of a sequence of network connections through consecutive hosts.

4.6 Conclusion

SPADE is a system for auditing, recording, and querying the provenance of distributed applications. Domain-specific (such as operating system level) activity is transformed into an OPM-compliant record by SPADEv2 modules. Each host maintains the authoritative repository of its data provenance. The distributed model of SPADE introduces the problem of reconstructing provenance during querying. SPADE uses novel provenance sketches to improve the performance of querying such provenance metadata. The provenance sketches determine which past network connections were relevant to a query, allowing all appropriate hosts to be contacted in parallel. Host-specific provenance subgraphs are computed using local transitive closures between network vertices. These subgraphs are retrieved in parallel and stitched together. SPADE is currently deployed within the NIGHTINGALE [30] project for auditing distributed workflows. Using real provenance data, we have shown the efficiency of SPADE's novel matrix filter summary data structure.

Acknowledgements. This material is based upon work supported by the National Science Foundation under Grant OCI-0722068 and IIS-1116414. Any opinions, findings, and conclusions or recommendations expressed in this material are those of the authors and do not necessarily reflect the views of the National Science Foundation.

References

1. Abraham, J., Brazier, P., Chebotko, A., Navarro, J., Piazza, A.: Distributed storage and querying techniques for a semantic web of scientific workflow provenance. In: IEEE International Conference on Services Computing (2010)
2. Bloom, B.: Space/time tradeoffs in hash coding with allowable errors. Communications of the ACM 13(7) (1970)
3. Broder, A., Mitzenmacher, M.: Network Applications of Bloom Filters: A Survey (2002)
4. Callahan, S., Freire, J., Santos, E., Scheidegger, C., Silva, C., Vo, H.: VisTrails: Visualization meets data management. In: ACM SIGMOD International Conference on Management of Data (2006)
5. Chang, F., Dean, J., Ghemawat, S., Hsieh, W., Wallach, D., Burrows, M., Chandra, T., Fikes, A., Gruber, R.: BigTable: A distributed storage system for structured data. In: 7th USENIX Symposium on Operating Systems Design and Implementation (2006)
6. Deelman, E., Blythe, J., Gil, Y., Kesselman, C., Mehta, G., Patil, S., Su, M., Vahi, K., Livny, M.: Pegasus: Mapping scientific workflows onto the Grid. Grid Computing (2004)
7. Dong, G., Libkin, L., Su, J., Wong, L.: Maintaining transitive closure of graphs in SQL. International Journal of Information Technology 5 (1999)
8. Foster, I., Vockler, J., Wilde, M., Zhao, Y.: Chimera: A virtual data system for representing, querying, and automating data derivation. In: 14th International Conference on Scientific and Statistical Database Management (2002)
9. Frew, J., Metzger, D., Slaughter, P.: Automatic capture and reconstruction of computational provenance. Concurrency and Computation 20(5) (2008)
10. Gadelha Jr., L., Clifford, B., Mattoso, M., Wilde, M., Foster, I.: Provenance management in Swift. Future Generation of Computer Systems 27(6) (2011)

11. Gehani, A., Lindqvist, U.: Bonsai: Balanced lineage authentication. In: 23rd Annual Computer Security Applications Conference. IEEE Computer Society (2007)
12. Gehani, A., Kim, M., Zhang, J.: Steps toward managing lineage metadata in Grid clusters. In: 1st USENIX Workshop on the Theory and Practice of Provenance (2009)
13. Gehani, A., Malik, T.: Efficient Querying of Distributed Provenance Stores. In: 8th Workshop on the Challenges of Large Applications in Distributed Environments (2010)
14. Groth, P.: Recording Provenance in Service-Oriented Architectures, Report, University of Southampton (2004)
15. Groth, P., Luck, M., Moreau, L.: A protocol for recording provenance in service-oriented grids. In: International Conference on Principles of Distributed Systems (2004)
16. Groth, P.: On the Record: Provenance in Large Scale, Open Distributed Systems. Thesis, University of Southampton (2005)
17. Groth, P.: A Distributed Algorithm for Determining the Provenance of Data, e-Science (2008)
18. Groth, P., Moreau, L.: Representing distributed systems using the Open Provenance Model. Future Generation Computer Systems 27(6) (2011)
19. H2, `http://www.h2database.com`
20. Heinis, T., Alonso, G.: Efficient lineage tracking for scientific workflows. In: ACM SIGMOD International Conference on Management of Data (2008)
21. Holland, D., Braun, U., Maclean, D., Muniswamy-Reddy, K., Seltzer, M.: Choosing a data model and query language for provenance. In: 2nd International Provenance and Annotation Workshop (2008)
22. Karvounarakis, G., Ives, Z., Tannen, V.: Querying data provenance. In: ACM SIGMOD International Conference on Management of Data (2010)
23. Karypis, G., Aggarwal, R., Kumar, V., Shekhar, S.: Multilevel hypergraph partitioning: Applications in VLSI domain. In: 34th Design and Automation Conference (1997)
24. Malik, T., Nistor, L., Gehani, A.: Tracking and sketching distributed data provenance. In: 6th IEEE International Conference on e-Science (2010)
25. Miles, S., Deelman, E., Groth, P., Vahi, K., Mehta, G., Moreau, L.: Connecting scientific data to scientific experiments with provenance. In: 3rd IEEE International Conference on e-Science and Grid Computing (2007)
26. Moreau, L., Ludaescher, B., Altintas, I., Barga, R., Bowers, S., Callahan, S., Chin Jr., G., Clifford, B., Cohen, S., Cohen-Boulakia, S., Davidson, S., Deelman, E., Digiampietri, L., Foster, I., Freire, J., Frew, J., Futrelle, J., Gibson, T., Gil, Y., Goble, C., Golbeck, J., Groth, P., Holland, D., Jiang, S., Kim, J., Koop, D., Krenek, A., McPhillips, T., Mehta, G., Miles, S., Metzger, D., Munroe, S., Myers, J., Plale, B., Podhorszki, N., Ratnakar, V., Santos, E., Scheidegger, C., Schuchardt, K., Seltzer, M., Simmhan, Y.: The First Provenance Challenge. Concurrency and Computation: Practice and Experience 20(5) (2007)
27. Moreau, L., Clifford, B., Freire, J., Futrelle, J., Gil, Y., Groth, P., Kwasnikowska, N., Miles, S., Missier, P., Myers, J., Plale, B., Simmhan, Y., Stephan, E., van den Bussche, J.: The Open Provenance Model core specification (v1.1). Future Generation Computer Systems (2010)
28. MySQL, `http://www.mysql.com`
29. Neo4j, `http://neo4j.org`
30. Novel Information Gathering and Harvesting Techniques for Intelligence in Global Autonomous Language Exploitation,
`http://www.speech.sri.com/projects/GALE/`
31. PlanetLab, `http://www.planet-lab.org`
32. Muniswamy-Reddy, K., Holland, D., Braun, U., Seltzer, M.: Provenance-aware storage systems. In: USENIX Annual Technical Conference (2006)

33. Muniswamy-Reddy, K., Braun, U., Holland, D., Macko, P., Maclean, D., Margo, D., Seltzer, M., Smogor, R.: Layering in provenance systems. In: USENIX Annual Technical Conference (2009)
34. Muniswamy-Reddy, K., Macko, P., Seltzer, M.: Making a Cloud provenance-aware. In: 1st USENIX Workshop on the Theory and Practice of Provenance (2009)
35. Muniswamy-Reddy, K., Macko, P., Seltzer, M.: Provenance for the Cloud. In: 8th USENIX Conference on File and Storage Technologies (2010)
36. Simmhan, Y.L., Plale, B., Gannon, D., Marru, S.: Performance evaluation of the Karma provenance framework for scientific workflows. In: 1st International Provenance and Annotation Workshop (2006)
37. Support for Provenance Auditing in Distributed Environments, `http://spade.csl.sri.com`
38. Speech Technology and Research, SRI International, `http://www.speech.sri.com`
39. Thain, D., Tannenbaum, T., Livny, M.: Condor and the Grid, Grid computing: Making the global infrastructure a reality. John Wiley (2003)
40. Tupelo project, NCSA, `http://tupeloproject.ncsa.uiuc.edu/node/2`
41. Zhou, W., Sherr, M., Tao, T., Li, X., Loo, B., Mao, Y.: Efficient querying and maintenance of network provenance at Internet-scale. In: ACM SIGMOD International Conference on Management of Data (2010)

Chapter 5
A Mobile Cloud with Trusted Data Provenance Services for Bioinformatics Research

Jinhui Yao, Jingyu Zhang, Shiping Chen, Chen Wang, David Levy, and Qing Liu

Abstract. Cloud computing provides a cheap yet reliable outsourcing model for anyone who needs large computing resources. Together with the Cloud, Service Oriented Architecture (SOA) allows the construction of scientific workflows to bring together various scientific computing tools offered as services in the Cloud, to answer complex research questions. In those scientific workflows, certain critical steps need the participation of research personnel or experts. It is highly desirable that scientists have easy access, such as mobile devices, to the workflows running in the Cloud. Furthermore, since the participants in this cross-domain collaboration barely trust each other, achieving reliable data provenance becomes a challenging task. This book chapter aims to discuss these issues and possible solutions. In this book chapter, we describe a Mobile Cloud system with a trusted provenance mechanism. The Mobile Cloud system facilitates the use of mobile devices to manipulate and interact with the scientific workflows running in the Cloud. Moreover, it provides trusted data provenance by acting as a trusted third party to record provenance data submitted by the participating services during the workflow execution. We have implemented a prototype which allows the bioinformatics workflow design and participation using mobile devices. We prove the concept of Mobile Cloud with the prototype and conducted performance evaluation for the significant points of bioinformatics workflow platform.

Jinhui Yao · Shiping Chen · Chen Wang
Information Engineering Laboratory, CSIRO ICT Centre, Australia
e-mail: {Jinhui.Yao, Shipping.Chen, Chen.Wang}@csiro.au

Qing Liu
Intelligent Sensing and Systems Laboratory, CSIRO ICT Centre, Australia
e-mail: q.liu@csiro.au

Jinhui Yao · Jingyu Zhang · David Levy
School of Electrical and Information Engineering, University of Sydney, Sydney, Australia
e-mail: {jin.yao, jingyu.zhang, david.levy}@sydney.edu.au

Q. Liu et al. (Eds.): Data Provenance and Data Management in eScience, SCI 426, pp. 109–128.
springerlink.com

5.1 Introduction

The emergence of computing resource provisioning known as the Cloud has revolutionized classical computing. It provides a cheap and yet reliable outsourcing model for anyone who needs large computing resources. Given the fact that many scientific breakthroughs need to be powered by advanced computing capabilities that help researchers manipulate and explore massive datasets [11], Cloud computing offers the promise of "democratizing" research, as a single researcher or small team can have access to the same large-scale compute resources as large as well-funded research organizations without the need to invest in purchasing or hosting their own physical IT infrastructure.

Together with the Cloud, the concept of Service Oriented Architecture (SOA) allows flexible and dynamic collaborations among different service providers. A service can either directly be used for its own functions or be composed with other services to form new value-added workflows [14]. Through SOA, scientific workflows can be used to bring together various scientific computing tools and compute resources offered as services in the Cloud to answer complex research questions. Workflows describe the relationship of individual computational components and their input and output data in a declarative way. In astronomy, for example, scientists are using workflows to generate science-grade mosaics of the sky [12], to examine the structure of galaxies [24]. In bioinformatics, researchers are using workflows to understand the underpinnings of complex diseases [17].

In the design of scientific workflows, certain critical steps need the participation of research personnel or experts. For example, details of the workflow design and which scientific tools need to be included must be decided by an expert in the area. Complex patterns generated from the experiments need to be visually inspected by the scientists who will, based on their domain knowledge and experience, determine the next steps for further analysis. In this regard, it is highly desirable that scientists have easy access, such as mobile devices, to the services and workflows running in the Cloud so that they can design and participate in the workflows efficiently at anytime anywhere.

Furthermore, data provenance has been widely acknowledged as an important issue for scientific experiments [31][22][19], for the provenance data collected during the experiment can be used to understand, reproduce the experiments conducted; identify the way data are derived. However, within this service-oriented collaboration, each service provider or individual researcher is from different companies, organizations or research institutions. The cross-domain collaboration intuitively suggests that the participants do not fully trust each other even though they need to collaborate. This implies they will question each other i) if a particular participant has employed proper data provenance mechanisms during the experiment; ii) if the recorded provenance data have been or will be tampered with; and iii) if the issuer and the integrity of the provenance data are somehow verifiable. These doubts caused by the lack of trustworthiness makes achieving reliable data provenance a challenging task. As a computing resource provider, these domain specific trust concerns are difficult for the Cloud infrastructure to address. Hence, these shortcomings reduce the incentives of individuals to participant in such cross-domain collaborative scientific workflow and are harmful for the wide adaptation of this computing paradigm. Therefore, a means to record the provenance data during the experiments in a trustworthy way is needed.

To address the above needs, in this book chapter, we describe a Mobile Cloud system with trusted provenance mechanisms. The Mobile Cloud system facilitates the use of mobile devices to manipulate and interact with the scientific workflows running in the Cloud. Through the light applications running in the mobile devices, the users can choose the services in the Cloud to form workflows for certain experiments, and be involved in the workflow execution, e.g. to conduct approval tasks as a step of the workflow. Mobile Cloud offers great mobility to the Cloud computing resources that, one can design, view, and participate in the scientific workflows running in the Cloud at anywhere and any time.

A significant aspect of the Mobile Cloud is its ability to provide trusted data provenance. The execution result of the workflows cross multiple domain service would be questionable without corresponding provenance data. In our system the Mobile Cloud serves as a trusted third party to record provenance data submitted by the participating services during the workflow execution. By enforcing strong accountability via the use of cryptographic techniques, the provenance data submitted by the participants in the workflow are undeniably linked to the submitter, which means its issuer and integrity can be cryptographically verified. With these verifiable provenance data recorded by a trusted platform, the collaborating entities can have a much better sense of trust in the validity of the provenance information they need to use. In general, the trusted data provenance implies that the provenance data collected is unarguably attributed to a particular service conducted the associated operations; and the chance any entity can tamper this data after it has been issued without being noticed is extremely low.

The main contributions of this book chapter are: 1) we design a Mobile Cloud system as a middleware layer to facilitate the use of mobile devices to design and interact with the scientific workflows running in the Cloud; 2) we define and illustrate the concept of strong accountability and the way it can be applied to record activity traces with provability; 3) we propose a novel approach to obtain activity traces from the execution of workflows and use them to construct data provenance graph to illustrate provenance information; and 4) we evaluate the performance of the Mobile Cloud system in the Cloud with real services.

5.2 The Application Scenario

In the area of gene research, the recent development of microarray technology [20] has led to rapid increase in the variety of available data and analytical tools. Some recent surveys published in *Nucleic Acids Research* show there are 1037 databases [4] and more than 1200 tools [2]. The analysis of microarray data commonly requires the biologists to query various online databases and perform a set of analysis using both local and online tools.

Following is an example. One of the important methods to understand colorectal cancer (CRC) is to understand the underlying molecular pathways involved in this disease. The rat azoxymethane (AOM) model of CRC is often used in dietary intervention studies as it induces mutations in genes which are also found to be mutated in human adenomas and adenocarcinomas. To define the baseline variation in global gene expression, the biologists extract RNA from mucosa scraped from colon and analyze the global gene expression using the Affymetrix Gene Chip. Data is normalized and then analyzed for differential expression. By contrasting the results from normal and cancer mice, biologists can identify candidate genes through statistical analysis. Further analysis – such as searching for the functions known to these genes are commonly performed to examine whether and how the candidate genes relate to the colorectal cancer. The followings are the data acquisition and analysis steps to perform the study of microarray experiment:

Quality control. It is to identify significant errors in the experiment, such as those caused by contaminated tissue samples. If any anomaly is detected by the biologists, the microarray result data are discarded.

Normalization. Microarray results from different samples need to be normalized before any meaningful comparison can be conducted.

Gene differentiation. By contrasting the results from cancerous and healthy tissues, differentially expressed genescandidate genes that are active in cancer are identified by applying some statistical methods (e.g. LIMMA).

Functional analysis. Most differentially expressed genes are further studied to understand the biological functions of the disease. There are various resources available for study. For example, gene symbols and descriptions could be retrieved from the Rat Genome Database and/or BioMart. Gene Ontology (GO) and KEGG databases could provide gene functions and molecular pathways information respectively. Experts need to be involved in order to make arrive at decisions as which study to conduct and which database to use.

We can see that the four standard analysis procedures listed above can not only be extremely computing intensive but also require some decision making from the research scientists or experts at certain critical steps (e.g. quality control). It clearly follows that, a viable approach to conducting such research must utilize a computing platform with enormous computing capacity, but which allows research scientists to interact easily with it. This is essentially the reason why we are promoting the "Mobile Cloud" - a composition of the Cloud and the mobile devices as a suitable paradigm for complicated bioinformatics research.

5.3 A Mobile Cloud System for Bioinformatics Research

As we have established in previous sections, the Mobile Cloud system composes the Cloud and the mobile devices to conduct complex bioinformatics research. The bioin-

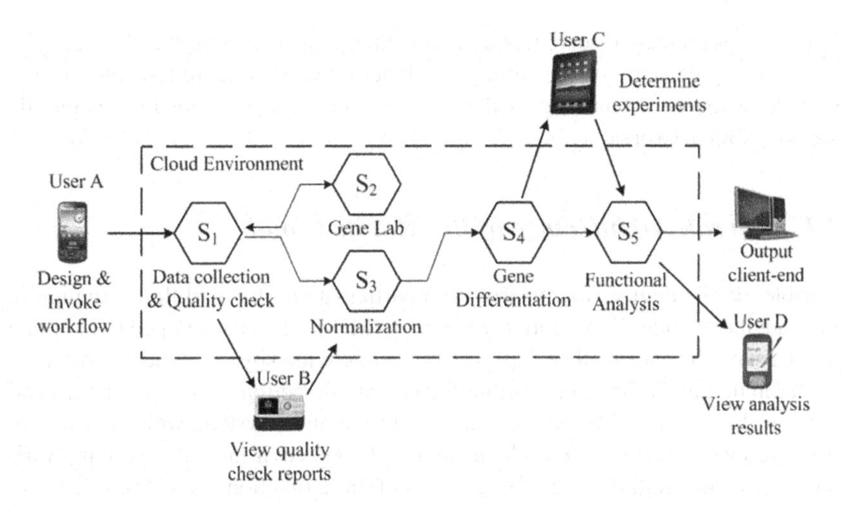

Fig. 5.1 Overview of the Mobile Cloud system

formatics research scenario we chose is the study of the cause of colorectal cancer (described in section 2); Fig. 5.1 shows the system with this research scenario.

In the Cloud, different computing-intensive gene-research tools are deployed by different research bodies and provided as services. Outside the Cloud, research scientists or gene analysts locate the desired services in the Cloud, and use them to compose a workflow for studying the cancer. In a gene research lab, we assume the gene data in the subject microarray chips are scanned and archived in some digital database, which can be reached from the Cloud or itself could be a Cloud storage service [18] such as Amazon S3. The Mobile Cloud operates as this: a researcher (user A) designs the scientific workflow and composes the needed services in the Cloud, then he invokes the first service "Data collection and Quality Check", which retrieves the gene data from the nominated "Gene Lab" where the gene subjects are stored, then conducts quality checks on the gene data. Once finished, the data is sent to the next service "Normalization" and a quality report is sent to user B for confirmation. If user B confirms the data quality, the normalization service will normalize the data and send the results to "Gene Differentiation". Another report is sent to user C After the differentiation, to choose the suitable experiment for the functional analysis. When the workflow is complete, the results are sent to a client end and a final report is sent to user D. We can see in the workflow multiple research scientists are involved. They participate in the workflow by using portable or desktop devices to invoke or receive output from the services.

Given the recent impressive advances in the mobile technology, the computing capability of mobile devices – however limited compared to desktops or laptops – is more than enough to run basic UI, display data sets and process reports. Then it is completely plausible that one can use some light applications running in his mobile device to design, execute and participate in the workflows running in the cloud that are composed of computing intensive scientific tools. The benefit of this is intuitive – scientists, researchers and any other people who need to leverage the vast computing capacity of the cloud to conduct scientific researches, can do so at anywhere in the world as long as there is internet connection for his/her mobile device (e.g. 3G network).

5.3.1 Overall Architecture of the Mobile Cloud

To enable mobile devices to construct and participate in the workflows running in the Cloud, the Mobile Cloud middleware layer (MC-layer) is developed to facilitate these. This middleware shall be deployed or even provided by the cloud environment provider in that environment to facilitate efficient interactions with the services and the clients. Fig. 5.2 provides an architectural design of the system, which consists of a user interface (residing on mobile devices), a Cloud environment containing various services and a middleware layer consists of three function units. The respective functionalities of its components are summarized as follows:

- *Cloud environment* provides various services deployed by respective providers and the MC-layer to facilitate the Mobile Cloud. The services have registered their access end-point with the MC-layer.
- *Service repository/composition* stores the information about the services in the Cloud that have registered with it. It helps the user to search for the services that best satisfy the requirements specified, and compose them into workflows.
- *Workflow execution* fulfils two functions: (a) orchestrating workflows during the operation; (b) invoking web services according to the defined workflow.
- *Trusted data provenance unit* (*TPU*) records cryptographically signed provenance data submitted by the participating services during the execution of the workflows. Using the recorded data, it monitors the status of the execution and allows the clients to query data provenance traces (this part will be elaborated in detail in section 4).
- *User interface* allows users to register, design workflows and participate in a running workflow.

For mobile devices to construct workflows, they first need to send a search request to the Service Repository in order to get a list of the services/workflows they are looking for. A convenient UI has been implemented on the mobile devices to allow the users easily to design the workflows using the services listed by the Service Repository (the UI will be elaborated in the evaluations). Once the workflow has been designed, a representative XML based description script is generated to be submitted to the Service Composition unit. The *Service Composition* unit thus according to the script, composes the services to form the desired workflows. The services can be composed in two ways: i) centrally composed, where the MC-layer invokes the

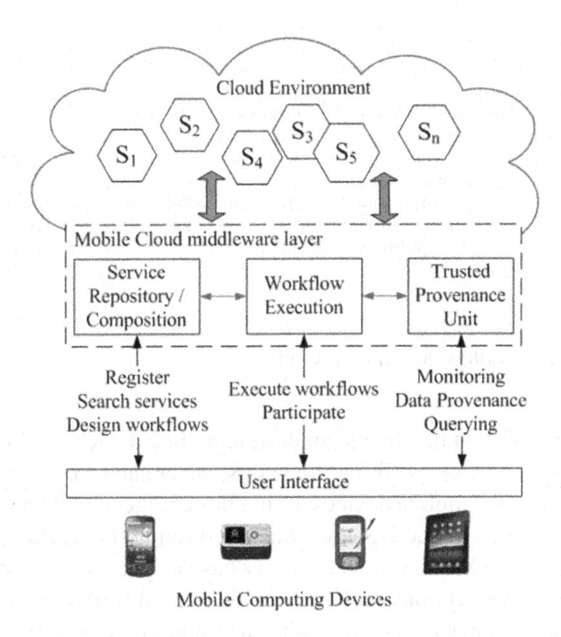

Fig. 5.2 Architecture of Mobile Cloud Middle layer

services in the sequence designed by the user; and ii) remotely orchestrated, where certain orchestration scripts such as BPEL [1] will be generated and distributed to all the services involved for deployment.

5.3.2 Workflow Design through Abstract Description Script

In our system, the workflow designed by the users is an abstract workflow, that is, the users only need to specify the type of service needed, and the MC-layer will search its service repository and recommend the best suited ones according to the user's specifications to let users select from. This enables researchers and scientists to use light weight editing applications in their mobile devices or desktop computers to write simple scripts to design the workflows. Listing 12 gives a sample of the abstract workflow description script. As it is developed based on the BPEL, "sequences" and "flows" are used to specify serial and parallel composition, and "Actions" are used to define the invocation operations. The sample describes the first half of the gene analysis workflow in Fig. 5.1. In some action, the endpoint is set to be "OPTIMAL". This is to tell the Service Composition unit to recommend the best suited services. When the abstract script is submitted to the MC-layer, a list of suitable services will be returned for selection.

```
1  <sequence name="main">
2      <Action operation="start" invoker="client" endpoint="QualityCheck"
3          type="send&forget"... />
4      <Action operation="fetchGene" invoker="QualityCheck"
5          endpoint="GeneLab" type ="send&receive"... />
6      <flow>
7          <Action operation="sendForApproval" invoker="QualityCheck"
                       endpoint="User B" type ="send&forget"... />
8          <Action operation="normalization" invoker="QualityCheck"
                       endpoint="OPTIMAL" type ="send&forget"... />
9      </flow>
10     ...
11 </sequence>
```

Listing 5.1 Sample workflow description script

As we have established in our system design, mobile devices will be involved in the workflows as web services. To facilitate this, we created a customized web service engine to run on the mobile devices. Using this engine, mobile devices can both send and receive service requests, as well as interpreting the workflow description scripts delivered by the MC-layer. Once a user has designed and submitted a workflow, the workflow description script will be forwarded to the researcher that are involved. The mobile devices they are using will interpret the workflow script and save the workflow logic. When a service request is received during the execution of the workflow, the UI will allow the user to view the content (e.g. quality check reports) and provide the list of the services that the user should send output request to according to the workflow logic (e.g. normalization services).

The MC-layer recommends services from its registry by matching the descriptive documentations of the services (e.g. WSDL, user annotations) with the requirements specified by the user. It provides a simple means for researchers who are not very familiar with the web service technologies or have not found the suitable service for a specific task. When all services involved in the workflow are determined, the orchestration engine in MC-layer will then execute the designed workflow. For the technical details of this aspect of the MC-layer, please refer to our previous publications about the Web Service Management System (WSMS) [26].

5.4 Accountability for Trusted Data Provenance

For scientific experiments, not only the resultant data are considered, the steps of how these data are derived along the process can also be very valuable. It has been widely realized that data provenance plays an important role in the scientific researches [21]. A mechanism is needed to preserve the intermediate data forms generated by different services and participating scientists during the execution of the workflow. However, the workflows in the Cloud are constructed using services provided by different parties who barely know each other, and the participating scientists are most likely from different institutions. The correctness of the resultant workflow relies on the individual correctness of all participators, that is, if the service or the individual is compliant to the pre-defined workflow logic, or Service

Level Agreement (SLA). The scientific integrity of the gene analysis results in our example scenario, will be highly questionable if the services and scientists involved, can act willy-nilly and get away with processing errors.

Therefore, trusted data provenance mechanisms are necessary in such systems with participants from different administrative domains. Provenance data should be preserved in a trustworthy way that, the contributors of the data are committed to their truthfulness. This naturally leads us to the issue of accountability.

5.4.1 Accountability for Trustworthiness

Accountability can be interpreted as the ability to have an entity account for its behaviors to some authorities [15]. This is achieved by binding each activity conducted to the identity of its actor with proper evidence [27]. Such binding should be achieved under the circumstance that all actors within the system are semi-trusted. That is, each identified actor may lie according to their own interest. Therefore, accountability should entail a certain level of stringency in order to maintain a system's trustworthiness. Below, we identify several desirable properties of a fully accountable system:

Verifiable: The correctness of the conducted process can be verified according to the actions and their bindings recorded.

Non-repudiable: Actions are bound to the actors through evidence, and this binding is provable and undeniable.

Tamper-evident: Any attempt to corrupt to recorded evidence inevitably involves the high risk of being detected.

Accountability can be incorporated into activity-based workflow by requiring the entity conducting the process to log non-disputable evidence about the activities in a separate entity. This is illustrated in Fig. 5.3. In the figure, after incorporating accountability into an ordinary process, entity A is now required to perform logging operations before and after conducting the activity in its process. The evidence is logged in a separate entity - entity B - so that entity A cannot access the logged evidence. The evidence needed to be logged should contain enough information to describe the conducting activity. In our simple example, which is intuitive enough, the evidence should include the states of the factors concerning the start of the activity (e.g. the input variables) and the factors concerning its completion (e.g. the output value and the parameters).

The logging operations require the employment of Public Key Infrastructure (PKI) in all involved service entities. Each of them has its own associated public-private key pair issued by certificated authorities which are used for signing purposes. The logging operations are as follows:

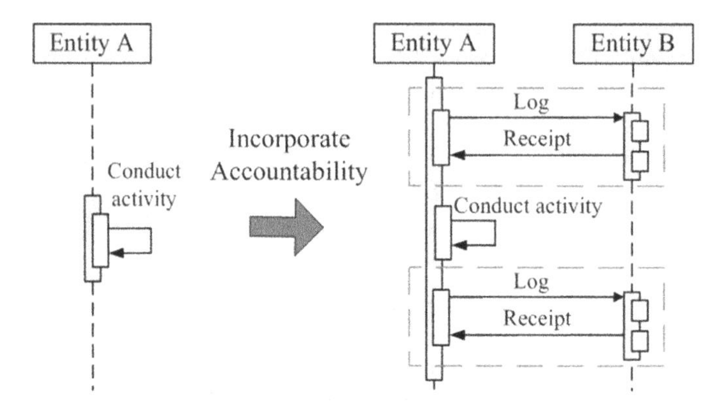

Fig. 5.3 Example of incorporating accountability into process

1. The logger - A signs the evidence E with his private key K_{a-} to create a digital signature of the evidence $\langle E \rangle_a$.
2. The evidence and its signature are then logged in a separate entity - B.
3. When received, B signs $\langle E \rangle_a$ with his private key to create a receipt $\langle \langle E \rangle_a \rangle_b$.
4. Lastly, the receipt is sent back to the logger in the reply.

Assuming the digital signature is un-forgeable. B can use $\langle E \rangle_a$ to prove A produced E. A can use $\langle \langle E \rangle_a \rangle_b$ to prove a separate entity - B has accepted $\langle E \rangle_a$ in the past, which is the evidence produced by A. With this logging procedure, both parties can prove the true issuer of the evidence logged, in our case, the provenance data produced.

5.4.2 Logging Provenance Data at Trusted Provenance Unit

In Mobile Cloud, the *Trusted Provenance Unit* (TPU) acts as the separate entity B, dedicated to provide accountability to all underlying services involved in the workflow. Fig. 5.4 shows the structure. All the mobile devices, service nodes in the Cloud as well as local computing nodes that are involved in the workflow, register with TPU and submit provenance data during the execution of the workflow.

The provenance data can be recorded in various ways, for instance, if the service invocations are all relayed by the MC-layer, they can be simply archived when received. Here we illustrate a generic approach to incorporate the data logging into the workflows by transforming the workflow descriptive scripts. Business process or workflows are often defined through process descriptive languages, which will be interpreted by orchestration engines (e.g. Apache ODE) to conduct the process accordingly. A good example of the process descriptive language is Business Process Execution Language (BPEL) [1]. BPEL models the business activities into several basic activity types, and then composes those types to describe the whole process. The core activity types include:

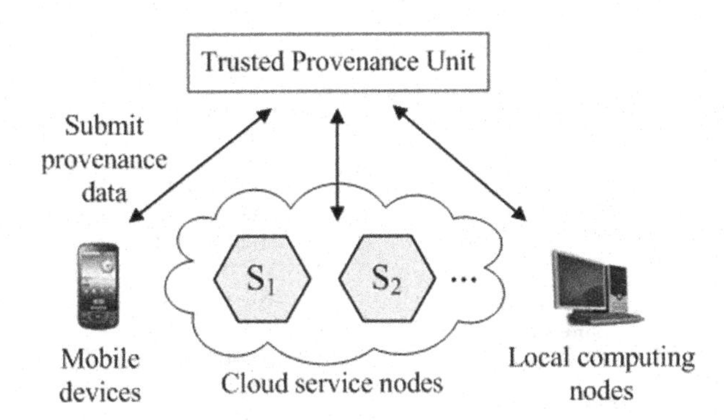

Fig. 5.4 TPU records provenance data from various sources

1. *Receive*, receiving the request from a requester. This activity type will specify the variable to which the input data is to be assigned.
2. *Invoke*, invocation to an endpoint (service). Invoke activity type will specify the variable used as the input and the variable used to store the output data for this invocation.
3. *Reply*, replying the invocation. A variable will be specified to be returned to the requester as the result.

To add logging activities into the workflow, we can insert *invoke* activity types into the BPEL script to invoke a certain endpoint (e.g. logging service) with the provenance data to be logged. And due to the distinct natures of *receive*, *invoke* and *reply* activity types, the rules used to decide the insertion locations are in fact quite straightforward. For the *receive* activity, an *invoke* should be inserted right after it, to log the input data received. For the *invoke* activity, one *invoke* should be inserted before this activity and another to be inserted after, to log the input data and the reply data of the invocation respectively. And finally for the *reply* activity, an *invoke* needs to be inserted just before it to log the result data that is about to be returned to the requester. The invocation endpoint for the *invoke* activities inserted (i.e. logging service) should either be a service in the same domain of the logger, or a trusted party nominated by the logger, which in turn signs the evidence on the logger's behalf and forward the signed evidence to the TPU.

To further illustrate this transformation process, we have presented an example in Fig. 5.5. Fig. 5.5a shows the graphical view of an ordinary sample BPEL. This simple process is started by receiving an input (ReceiveInput); then a partner link (collaborating service) is invoked in turn (InvokePartnerLink), and finally, replies the result to the client (ReplyClient). Fig. 5.5b is the BPEL after the transformation. We can see in Fig. 5.5b that four logging invoke activities (the InvokeLogging) have been inserted, one after the "ReceiveInput"; one before and one after "InvokePartnerLink"; and one before "ReplyClient". Because BPEL is entirely based on xml schema, any xml schema parser will be capable of analyzing and inserting activities into it. The implementation details of the incorporation of accountability have been elaborated in our previous work [25].

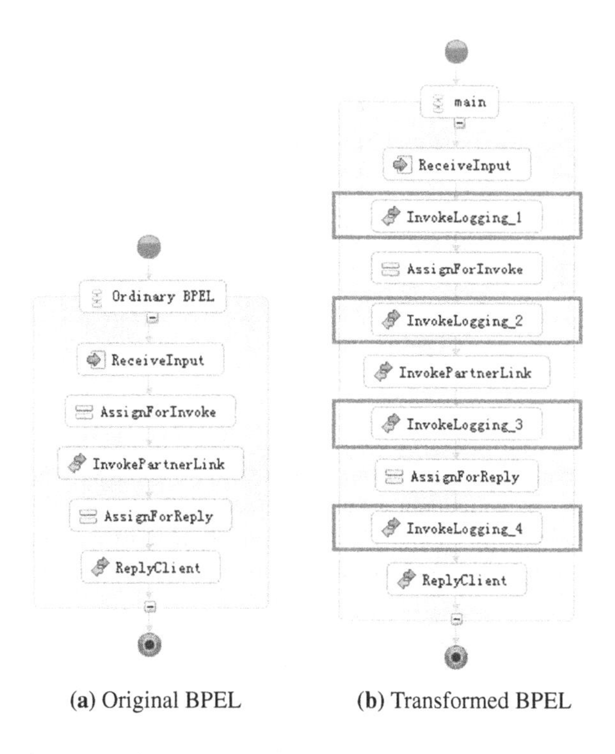

(a) Original BPEL **(b)** Transformed BPEL

Fig. 5.5 Transformation of BPEL

5.4.3 Architectural Design of Trusted Provenance Unit

TPU is responsible of recording the provenance data from all the participants of the workflow. As accountability requires the submitter of the data to sign the data before submission to commit its truthfulness, services and the entities involved in the workflow are needed to register their identity documents (e.g. X.509) at MC-layer. When a new abstract workflow is proposed by a researcher, the Service Repository/Composition unit first find the services that best suit the specified requirements, then the filled workflow script is transformed by TPU to have logging activities (refer to Chapter 6 for details). Meanwhile, TPU uses the knowledge obtained from the documents registered to generate analysis logics to process the incoming data during the execution of the workflow. The resultant data provenance information will be delivered to the user through querying and visual displays.

The internal architectural design of TPU is shown in Fig. 6. In the initialization phase, registered information about the services, like WSDL, X.509 certificate etc.; and information about the workflows, like BPEL scripts are transmitted to TPU from Service Repository/Composition unit. TPU first transform the workflow script to incorporate logging activities and send the transformed script for redeployment; then it uses the registration information received to generate two components: "Monitoring Logic" and "Provenance Logic". In the monitoring phase, the provenance data

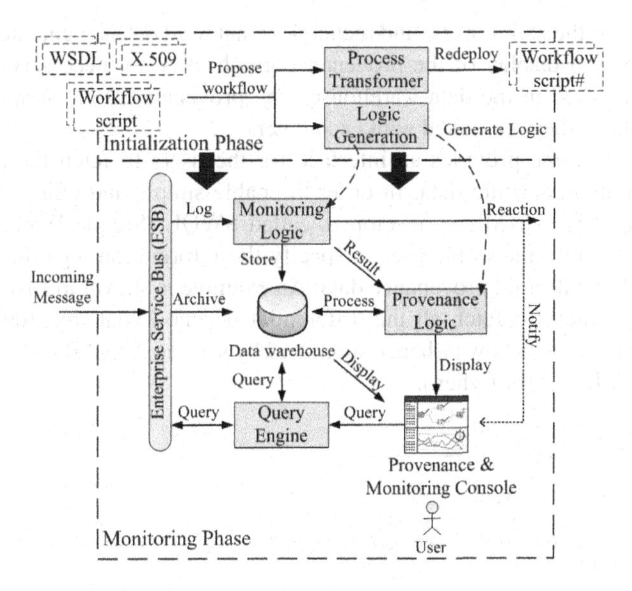

Fig. 5.6 Internal architecture of *Trusted Provenance Unit*

will be submitted from the participants in the workflow. These data will first be analysed by the monitoring logic to find obvious compliance violations (e.g. QoS service level agreement); then be processed by the provenance logic to generate data provenance information to be stored in the data warehouse.

When the provenance data are received, the provenance logic first labels the provenance data with information regarding the "four Ws": who, when, where and what. In general, the provenance logic will add labels explaining what this provenance data is about, in which workflow (where) it is generated, at what time and by which participant (who). Then, based on the knowledge obtained from the documentation registered, the provenance logic links the different provenance data with Open Provenance Model [13] edges to form a provenance graph. An example of such graph is displayed in Fig. 5.7. We can see from the example, the provenance information of the data pieces (circles marked with numbers) are expressed in terms

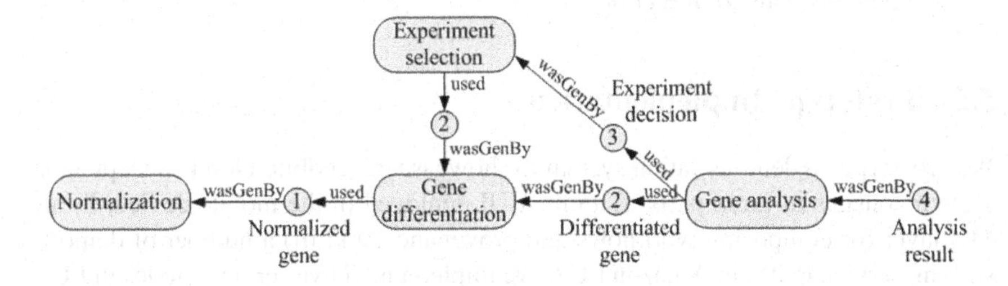

Fig. 5.7 An example of provenance graph

of their links to the activities (round rectangles) that used or/and generate them. The figure is a visual display of the provenance graph, it is not necessarily an actual graph when stored in the data warehouse. The provenance logic simply needs to label the data so they are linked with each other.

The query engine provides an interface for the users to fetch the provenance information about specific data. In order to enable simple and efficient querying, a query language in XML is developed, called SWQL (Simple Workflow Query Language). SWQL allows the user to specify the information regarding the "four Ws" to fetch the desired provenance data. An example is shown in Listing 16. The example is a query to fetch all the differentiated gene (what) recorded from the colorectal cancer workflow (where), submitted by service A and B (who) from 9am to 5pm on 20 July 2011 (when).

```
 1 <SWQL>
 2     <Action>Find</Action>
 3     <DataIdentifier>
 4         <Type>Differentiated gene</Type>
 5     </DataIdentifier>
 6     <EntityIdentifier>
 7         <Entity>Differentiation service A</Entity>
 8         <Entity>Differentiation service B</Entity>
 9     </EntityIdentifier>
10     <TimeInterval>
11         <From>9AM-20JUL2011</From><To>5PM-20JUL2011</To>
12     </TimeInterval>
13     <WorkflowIdentifier>Colorectal cancer</WorkflowIdentifier>
14     ...
15 </SWQL>
```

Listing 5.2 An example of SWQL query

The provenance and monitoring console is a graphical user interface to display provenance and monitoring information as well as let users query the data warehouse. During the execution of the workflow, the evolution of the data will be displayed in terms of the provenance graph generated by the provenance logic, and the status of the workflow will been shown. More details about the console will be discussed in the evaluation section.

5.5 Prototype Implementation

We prototyped a demonstration system to showcase our mobile-cloud concept. Our system consists of three parts: i) a client UI deployed in the mobile device; ii) an MC-layer for composing workflows and provenance; and iii) a number of demonstrating service nodes in Amazon EC2. We implemented five services nodes in EC2 to represent the gene research tools provided by different organizations. The services are linearly composed (one node finishes its job then invokes the next) to form a workflow using BPEL. The information about the services as well as the workflow are registered at the MC-layer, which is deployed in another computing instance in EC2. A remote user designs and invokes the workflow using the client UI locally deployed in the mobile device. With this setting, in this section, we will elaborate the implementation of the client UI; examine the communication overhead introduced when provenance data are logged at TPU during the execution ; and we show some processing latency when a real gene database (KEGG) is involved in a workflow.

The UI on mobile device is developed using Java platform, micro edition (J2ME). The mobile web service feature is deployed and runs on a HTC 9500 mobile phone, which is running on IBM Websphere Everyplace Micro Environment that supports a connected device configuration (CDC1.1). Fig. 5.8a and 5.8b show two screen shots of the Mobile Gene Management System (MGMS) - a scientific workflows design and surveillance tools. A user can define or edit a scientific process from the "New Work" button or "Previous Work" button as shown in Fig. 5.8a. Then, the user can select into process items and specify their detail information as shown in Fig. 5.8b. System users define the steps from four aspects, what services carry out these tasks; the number of child nodes; which methods/services are invoked; and what are the inputs and outputs of each step. Finally, an abstract workflow in BPEL will be generated and uploaded to the WSMS in Cloud, which will instantiate the abstract workflow by filling up the endpoints in the BPEL with the best concrete services URLs.

We have conducted testing to evaluate the latency introduced by incorporating the logging actions into the workflow. Figure 5.9a shows the overall latency to finish the process with untransformed BPEL scripts and with transformed ones. We have tested the workflow with request message size from 0.1KB (equivalent to a sentence) to 50KB (equivalent to a medium size document). For the process with transformed BPEL scripts to log the entire input/output messages (the series marked with "circles"), the latency introduced compared to the untransformed one (the series marked with "squares") grows as the request message becomes larger. In percentage terms, on average we observed a 30% increase in the overall process latency.

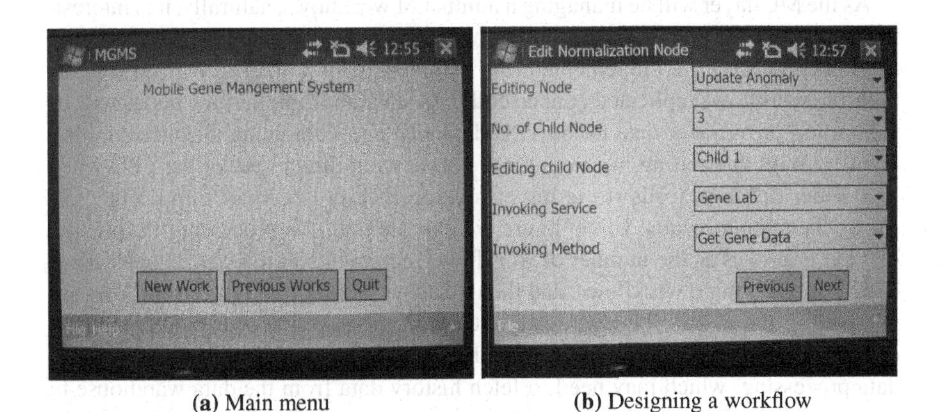

(a) Main menu **(b)** Designing a workflow

Fig. 5.8 Screen shots of Mobile Cloud client end UI

Intuitively, this latency is significant to the business process; however it can be improved through the use of hash functions. We can see in the graph, the extra latency is significantly reduced if the BPEL scripts are transformed only to log the hash of the evidence (the series marked with "triangles"). In fact, the extra latency almost remains constant regardless of the size of the request message, so it becomes more and more negligible when the message size increases.

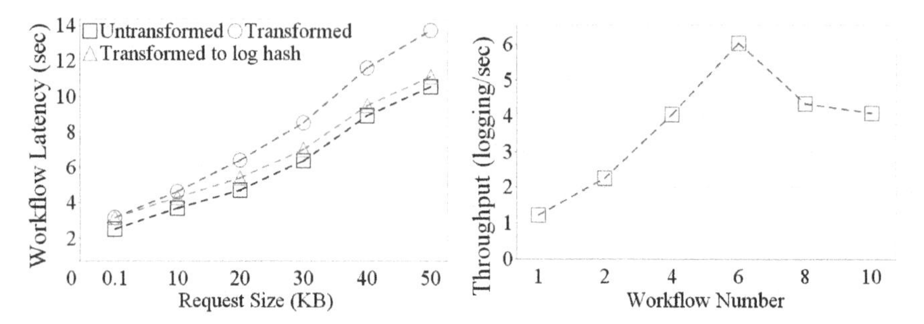

(**a**) Overall execution latency of the workflow (**b**) Throughput of TPU under different load

Fig. 5.9 Performance evaluation

In practice, it is very rare that the entire communication message is urgently needed to be logged at runtime for provenance purposes. Instead, the hash of the message computed using collision-resistant hash functions (e.g., SHA-1), which is a very small digest (160 bits for SHA-1), can be logged as a substitute. Because the hashes computed are collision-resistant, which means it is theoretically impossible to have two different items with the same hash, so the hash can be logged to represent the data. When the system is idle, the provenance data can be eventually logged and verified according to the hash values.

As the MC-layer will be managing a number of workflows, naturally, it is interesting to find out the processing capability of the TPU. To evaluate this, we replicated the workflow we have implemented (the colorectal cancer workflow), and execute multiple workflows replicated concurrently. As such, multiple service nodes will be submitting provenance data to the TPU deployed in a computing instance simultaneously. With this setting, we evaluate the processing throughput of the TPU when it is under different loads (in terms of logging received per unit time). Fig. 5.9b shows the testing results. In the figure we can see that, the processing throughput of TPU improves as the number of workflows increments, it reaches its peak when TPU is monitoring 6 workflows, and then it decays gradually if more workflows are involved in the monitoring. We tested this with messages of size 50KB, the processing operations conducted by AS involves both SLA monitoring and provenance data processing, which may need to fetch history data from the data warehouse to make conclusions. Since the computing power of a computing instance is fixed, an decrease in message size or processing complexity will shift the peak towards right to occur when more workflows are involved, and vice versa.

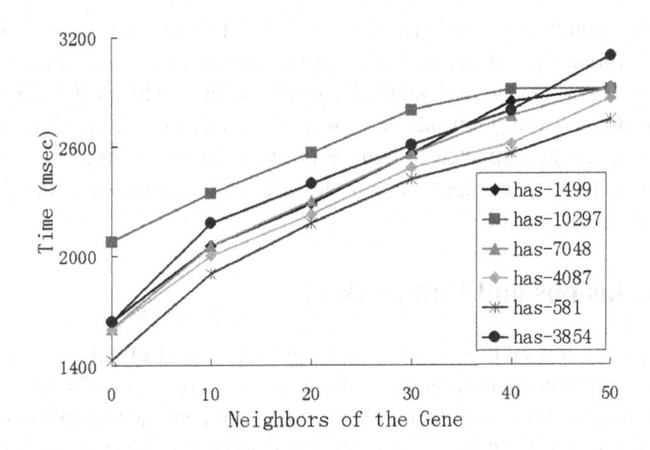

Fig. 5.10 Gene retrieval experiment with KEGG

To evaluate the performance of gene retrieving from gene bank services, we selected 6 example genes which are the genetic causes of colorectal cancer and retrieve their genetic neighbors from KEGG disease Database [8]. We test the response time from 0 neighbors to 50 neighbors. As shown in Fig. 5.10, it is clear that the latency is slowly increasing with changing the number of neighbors. The has-581 continually kept the best performance at all stages from the 1427msec for retrieving 0 gene neighbor to 2746.8msec for getting 50 neighbors. However, has-10297 spent 2078msec to search 0 neighbors and it cost 2912.6msec for finding 50 neighbors.

5.6 Related Work

Mobile computing provides a luggable computation model for users. Its portability makes it very ideal for many application scenarios. To extend its limited computing power, research communities have proposed novel designs to leverage the Cloud. [7] proposed a virtual cloud system, [28] detailed a distributed computing platform using mobile phones. They improve the capacities of mobile phones in the purpose of storage and computation. [5][10][9] presented some computation offloading studies that move some parts of the applications to run on the Cloud. Executing parts of application remotely can save battery lifetimes and significantly extend computing resources. However, these solutions do not support platform-independent cooperative interaction over an open network. In addition, after moving some parts of applications from stand-alone handheld devices to the cloud, several issues need to be considered in advance such as privacy, trustworthy or provenance.

The importance of provenance for scientific workflows has been widely acknowledged by various research communities. Many approaches have been proposed to record the derivations of the data during the scientific process. Approaches like [3][30] allow the designer to capture the intermediate data forms generated by the experiments at different granularities. In our work, we introduced the concept of accountability which not only provides trusted data provenance but can enforce compliance among the service providers. Compliance assurance has been studied decently in recent years, some remarkable works include [16][6][29][23]. Our work differs from them at the point that we consider a more hostile environment where all service entities are expected to behave in any possible manner and deceive for their own benefit. Cryptographic techniques are deployed in our system to ensure the evidence are undeniable.

5.7 Conclusions and Future Work

Cloud computing has emerged as a way to provide a cost effective computing infrastructure for anyone with large needs for computing resources. Together with the Service Oriented Architecture, research scientists can construct scientific workflows composed of various scientific computing tools offered as services in the Cloud to answer complex research questions.

In this book chapter, we have described a Mobile Cloud system which enables mobile devices to design and participate in the scientific workflows running in the Cloud. The scientific researchers can use mobile devices to sketch an abstract workflow design to be submitted to the mobile cloud middleware layer, which will recommend and compose the optimal services according to the designer's requirements. On top of that, we further incorporated accountability mechanisms to provide trusted data provenance during the execution of the scientific workflows. Trusted data provenance implies that the recorded provenance data about a certain workflow is cryptographically verifiable to be attributed to the responsible services who, issued them. The provenance data thus can be used with confidence that its source is verifiable and its integrity has been preserved.

In the future development, it will be interesting to explore the utilization of the trusted provenance data collected, to improve the service recommendation for workflow design. The applicability of a particular service in a certain workflow and its performances in the past executions can provide much information to the research scientists and the recommendation system about characteristics of this service and its eligibility for the workflow under design. Another direction of development is to utilize existing workflow platforms or service repositories (e.g. BioCatalogue[1]) to construct workflows and provide trusted data provenance. In this way we can testify the concept of Mobile Cloud and trusted data provenance in the practise, improve our methodology so as to offer more value and insights to the community.

References

1. Andrews, T., Curbera, F., Dholakia, H., et al.: Business process execution language for web services (BPEL4WS) specifications (2003), `http://download.boulder.ibm.com/ibmdl/pub/software/dw/specs/ws-bpel/ws-bpel.pdf`

[1] BioCatalogue, life science web service registry: http://www.biocatalogue.org/

2. Brazas, M., Fox, J., Brown, T., McMillan, S., Ouellette, B.: Keeping pace with the data: 2008 update on the bioinformatics links directory. Nucleic Acids Research 36, W2–W4 (2008)
3. Foster, I., Vockler, J., Wilde, M., Zhao, Y.: Chimera: A virtual data system for representing, querying, and automating data derivation. In: Scientific and Statistical Database Management Conference (2002)
4. Galperin, M.: The molecular biology database collection: 2008 update. Nucleic Acids Research 36, D2–D4 (2007)
5. Giurgiu, I., Riva, O., Juric, D., Krivulev, I., Alonso, G.: Calling the cloud: enabling mobile phones as interfaces to cloud applications. In: ACM/IFIP/USENIX International Conference on Middleware, pp. 83–102 (2009)
6. Huang, M., Peterson, L., Bavier, A.: Planetflow:maintaining accountability for network services. In: ACM SIGOPS Operating Systems Review, pp. 89–94 (2006)
7. Huerta-Canepa, G., Lee, D.: A virtual cloud computing provider for mobile devices. In: ACM Workshop on Mobile Cloud Computing Services: Social Networks and Beyond, San Francisco, pp. 61–65 (2010)
8. Kanehisa, M., Goto, S., Furumichi, M., Tanable, M., Hirakawa, M.: Kegg for representation and analysis of molecular networks involving diseases and drugs. Nucleic Acids Research 38, 355–360 (2010)
9. Kemp, R., Palmer, N., Kielmann, T., Bal, H.: Cuckoo: a computation offloading framework for smartphones. In: International Conference on Mobile Computing, Applications, and Services, pp. 62–81 (2010)
10. Kumar, K., Lu, Y.: Cloud computing for mobile users. Journal of Computer 18(99), 51–56 (2010)
11. Lu, W., Jackson, J., Barga, R.: Azureblast: A case study of developing science. In: Workshop on Scientific Cloud Computing, pp. 413–420 (2010)
12. Montage, http://montage.ipac.caltech.edu
13. Moreau, L., Freire, J., Futrelle, J., McGrath, R., Myers, J., Paulson, P.: The open provenance model (2007), http://eprints.ecs.soton.ac.uk/14979
14. Moser, O., Rosenberg, F., Dustdar, S.: Non-intrusive monitoring and service adaptation for ws-bpel. In: International Conference on World Wide Web, pp. 815–824. ACM, New York (2008)
15. Mulgan, R.: Accountability: An ever-expanding concept? In: Public Administration, pp. 555–573 (2000)
16. Mulo, E., Zdun, U., Dustdar, S.: Monitoring web service event trails for business compliance. In: International Conference on Service-Oriented Computing and Applications, pp. 1–8 (2009)
17. Oinn, T., Li, P., Kell, D., Goble, C., Goderis, A., Greenwood, M., Hull, D., Stevens, R., Turi, D., Zhao, J.: Taverna/myGrid: Aligning a workflow system with the life sciences community. Workflows in e-Science. Springer (2006)
18. Palankar, M., Iamnitchi, A., Ripeanu, M., Garfinkel, S.: Amazon s3 for science grids: a viable solution. In: Workshop on Data-Aware Distributed Computing, pp. 55–64 (2008)
19. Scheidegger, C., Koop, D., Santos, E., Callahan, H., Freire, J., Silva, C.: Tackling the provenance challenge one layer at a time. Concurrency and Computation: Practice and Experience 20(5), 473–483 (2008)
20. Schena, M., Shalon, D., Davis, R., Brown, P.: Quantitative monitoring of gene expression patterns with a complementary dna microarray. Science 270, 467–470 (1995)
21. Simmhan, Y., Plale, B., Gannon, D.: A survey of data provenance in e-science. ACM SIGMOD Record 34(3), 31–36 (2005)

22. Simmhan, Y., Plale, B., Gannon, D.: Karma2: Provenance management for data driven workflow. International Journal of Web Services Research 5(2), 1–22 (2008)
23. Squicciarini, A., Lee, W., Thuraisingham, B., Bertino, E.: End-to-end accountability in grid computing systems for coalition information sharing. In: Workshop on Cyber Security and Information Intelligence Research (2008)
24. Taylor, I., Shields, M., Wang, I., Philp, R.: Distributed p2p computing within triana: A galaxy visualization test case. In: IEEE International Parallel and Distributed Processings Symposium (2003)
25. Yao, J., Chen, S., Wang, C., Levy, D., Zic, J.: Accountability as a service for the cloud. In: IEEE International Conference on Services Computing, pp. 81–90 (2010)
26. Yu, Q., Liu, X., Bouguettaya, A., Medjahed, B.: Deploying and managing web services: issues, solutions, and directions. The VLDB Journal 17, 537–572 (2008)
27. Yumerefendi, A.R., Chase, J.S.: Strong accountability for network storage. ACM Trans. Storage 3(3), 11 (2007), doi: http://doi.acm.org/10.1145/1288783.1288786
28. Zhang, J., Levy, D., Chen, S., Zic, J.: mbosss+: A mobile web services framework. In: IEEE Asia-Pacific Services Computing Conference, pp. 91–96 (2010)
29. Zhang, Y., Lin, K.J., Hsu, J.: Accountability monitoring and reasoning in service-oriented architectures. Service Oriented Computing and Applications 1, 35–50 (2007), doi:10.1007/s11761-007-0001-4
30. Zhao, J., Goble, C., Stevens, R., Bechhoffer, S.: Semantically linking and browsing provenance logs for escience. In: International Conference on Semantics of a Networked World (2004)
31. Zhao, J., Goble, C., Stevens, R., Turi, D.: Mining taverna's semantic web of provenance. Concurrency and Computation: Practice and Experience 20(5), 463–472 (2008)

Chapter 6
Data Provenance and Management in Radio Astronomy: A Stream Computing Approach

Mahmoud S. Mahmoud, Andrew Ensor, Alain Biem, Bruce Elmegreen, and Sergei Gulyaev

Abstract. New approaches for data provenance and data management (DPDM) are required for mega science projects like the Square Kilometer Array, characterized by extremely large data volume and intense data rates, therefore demanding innovative and highly efficient computational paradigms. In this context, we explore a stream-computing approach with the emphasis on the use of accelerators. In particular, we make use of a new generation of high performance stream-based parallelization middleware known as InfoSphere Streams. Its viability for managing and ensuring interoperability and integrity of signal processing data pipelines is demonstrated in radio astronomy.

IBM InfoSphere Streams embraces the stream-computing paradigm. It is a shift from conventional data mining techniques (involving analysis of existing data from databases) towards real-time analytic processing. We discuss using InfoSphere Streams for effective DPDM in radio astronomy and propose a way in which InfoSphere Streams can be utilized for large antennae arrays. We present a case-study: the InfoSphere Streams implementation of an autocorrelating spectrometer, and using this example we discuss the advantages of the stream-computing approach and the utilization of hardware accelerators.

6.1 Introduction

Started in the 1930s, radio astronomy has produced some of the greatest discoveries and technology innovations of the 20th century. One of these innovations – radio interferometry and aperture synthesis – was awarded a Nobel Prize for Physics in

Mahmoud S. Mahmoud · Andrew Ensor · Sergei Gulyaev
AUT Institute for Radio Astronomy & Space Research, Auckland NZ
e-mail: {mmahmoud, aensor, sgulyaev}@aut.ac.nz

Alain Biem · Bruce Elemgreen
IBM T J Watson Research Center, Yorktown Heights NY
e-mail: {abiem, bge}@ibm.com

Q. Liu et al. (Eds.): Data Provenance and Data Management in eScience, SCI 426, pp. 129–156.
springerlink.com

1974 (Martin Ryle and Antony Hewish, 1974). An aperture synthesis radio telescope consists of multiple receiving elements in an array that observe the same radiating source(s) simultaneously. Essentially, an array of radio telescopes is used to emulate a much large telescope with size that of the diameter of the array, enabling a better angular resolution of the radio source(s) to be obtained. While angular resolution is determined by the array diameter, another important characteristic is telescope sensitivity, which is determined by its collecting area. The Square Kilometer Array (SKA) will be an aperture synthesis radio telescope, scheduled for completion in the 2020s, that will combine both factors, resolution and sensitivity. The total SKA collecting area of one square kilometer (10^6 m^2) will provide sensitivity that is 50-100 times higher than that of the best current radio telescope arrays. Its high angular resolution will be provided by distributing the square kilometer of collecting area into many stations that are spread out on a continental scale (with the baseline between some antennae over 3000 km).

A radio telescope antenna element detects electromagnetic waves by a current induced in an antenna receiver system. This can be measured as a voltage $s_i(t)$ at receiver i that is sampled and digitized at regular times t. Whereas a single receiver can measure the source brightness $I(\mathbf{d})$ in a specific direction $\mathbf{d}$, a pair of receivers i, j separated by a baseline vector $\mathbf{B}_{ij}$ can be used as an interferometer to measure the difference in phase between the signals s_i and s_j due to the time delay $\tau_{ij} = \mathbf{B}_{ij} \cdot \mathbf{d}/c$ between the received signals as illustrated in Figure 6.1.

The time delay τ_{ij} can be roughly approximated by the geometry of the antennae relative to the source direction (provided by an *Ephemerides service*), and more precisely determined by the resulting interference pattern in the cross-correlation between the signals:

$$(s_i \star s_j)(\tau) = \int_{-\infty}^{\infty} \overline{s_i(t)} s_j(t+\tau)\, dt.$$

The value $V_{ij} = (s_i \star s_j)(\tau_{ij})$ is termed a *visibility* and gives a source brightness measurement at a (u, v) point in the Fourier domain determined by the baseline $\mathbf{B}_{ij}$ for that pair of antennae. An array of n antennae has $\frac{n(n-1)}{2}$ baselines (one per pair of antennae) and so $\frac{n(n-1)}{2}$ visibilities can be obtained. However, if readings are taken over an interval of hours then each baseline changes over time due to the rotation of the Earth. Hence, over time the baselines sweep out elliptical arcs in the Fourier plane, illustrated in Figure 6.2.

The Fourier domain coverage of an array is the combination of the (u, v) tracks from all baselines provided by the array. It shows where the array samples the Fourier transform of the source image. For high quality imaging, it is desirable to have the best possible coverage of the Fourier domain, which is effectively the telescope aperture. A perfect source brightness distribution (the image of the radio source) could be obtained simply by taking the inverse Fourier transform if all (u, v) points in the Fourier domain were able to be measured, but this is never the case. A deconvolution process such as *Clean* or *Maximum Entropy* used in any modern

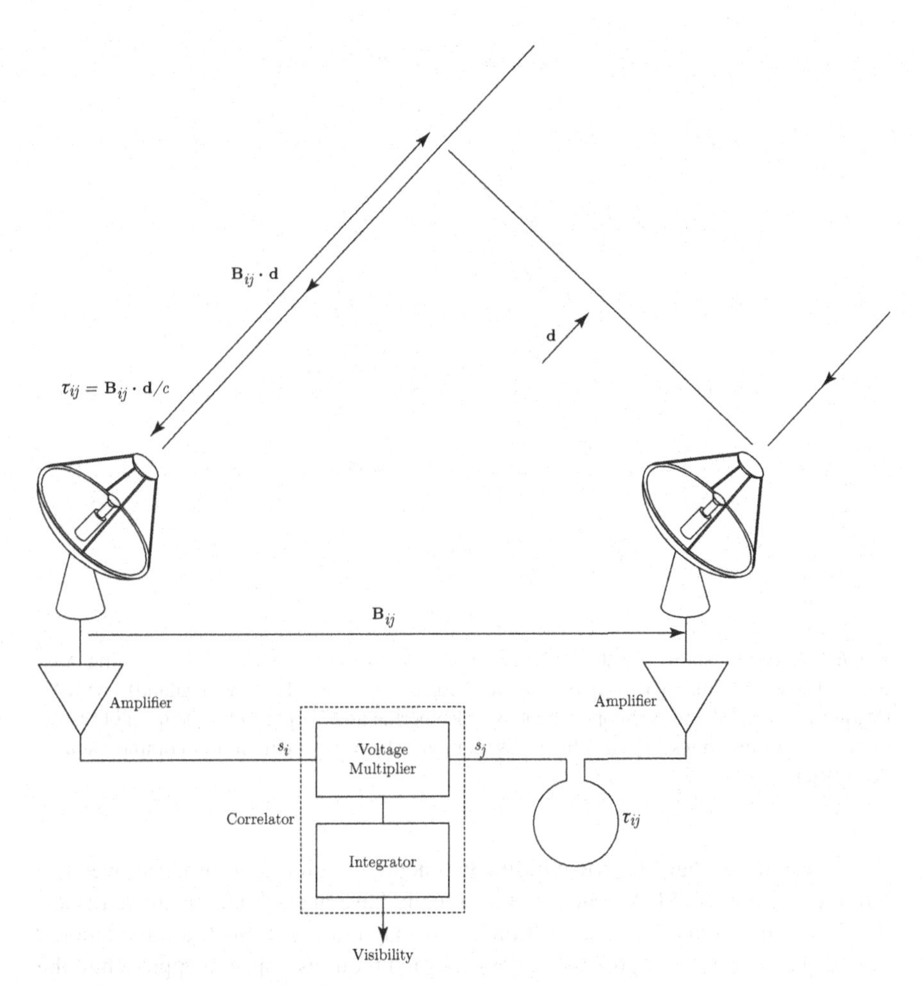

Fig. 6.1 Two-element interferometer

interferometer imaging can be thought of as a scheme for interpolating or extrapolating from the measured (u, v) points to all other points in the (u, v) plane [1].

Measuring the signals from all n antennae over a period of hours results in an enormous dataset for a large array and its processing is a very compute intensive problem. Figure 6.3 shows the operations that are performed in an array on the digitized signals in a simplified pipeline from raw data through to analyzed data products. Computational power required for these operations can be very significant, particularly for the Correlation operation which calculates the visibility V_{ij} for each of the $\frac{n(n-1)}{2}$ baselines at each time t via cross-correlations. It also performs an autocorrelation of each signal as discussed in Section 6.4, which together with the visibilities forms the *datacube* for the array at time t.

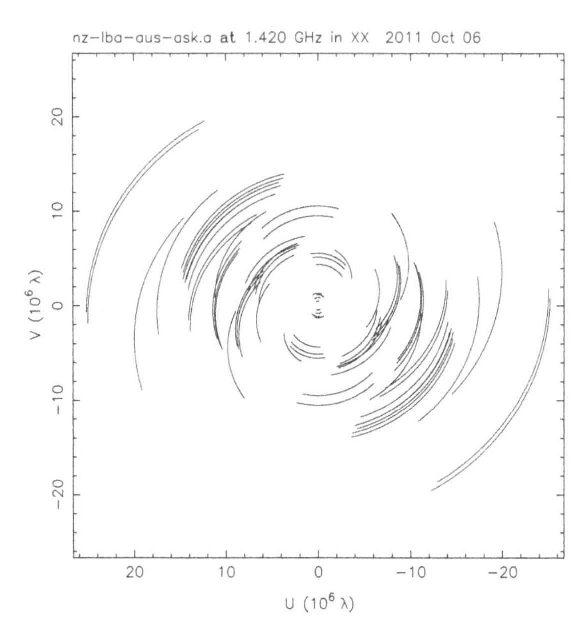

Fig. 6.2 A uv-plot shown for the VLBI array in Australia and New Zealand, including Australian Long Baseline Array (five radio telescopes), Warkworth (New Zealand), ASKAP (Western Australia) and AuScope antennas in Katherine and Yarragadee. The uv-plot results in 36×2 baseline tracks for a 4 hour observation, 21-cm wavelength and common source declination of $-70°$ [2].

It is estimated that LOFAR with its 36 antennae stations can produce over 100 TB/day [3]. For the SKA which will eventually have about 3000 antennae dishes, the data will increase by at least 5 orders of magnitude [4]. Such a huge amount of data places very high processing demands and requires a special approach to the overall organization of how data are processed and stored. It is only feasible to store the digitized raw signals required for calculating data cubes for small arrays and is limited to measurements taken over short time periods; in all other cases the data storage requirements are too large to be practical.

In the next section we introduce the stream computing paradigm and how IBM's InfoSphere Streams data management middleware utilizes this paradigm. In Section 6.2 we describe how InfoSphere Streams can be applied to the operational facets of large radio astronomy telescope arrays to handle the enormous data volumes and compute intensive operations. In Section 6.4 we consider an actual InfoSphere Streams application that performs streaming autocorrelations of actual radio astronomical observations.

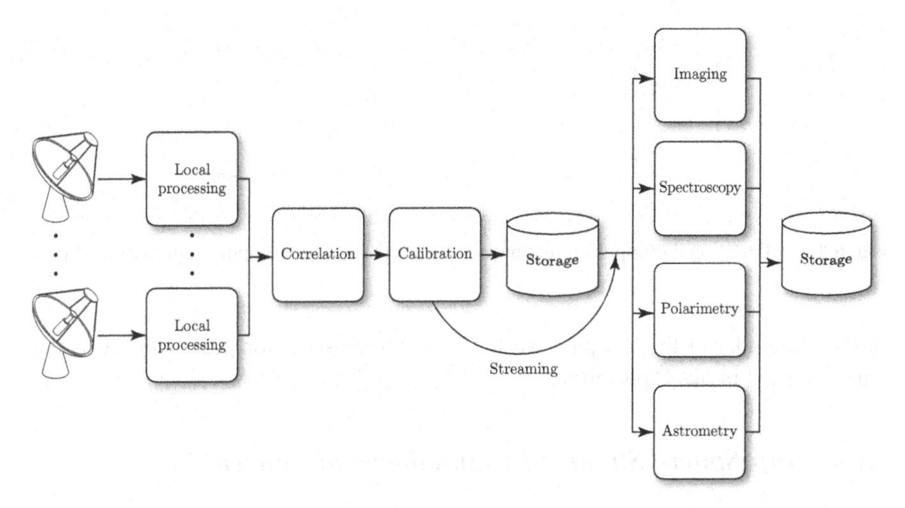

Fig. 6.3 Radio astronomy array pipeline

6.2 IBM InfoSphere Streams and the Stream-Computing Paradigm

With the vast expansion of data volumes generated in the current information age, there has been a paradigm shift in data management toward the processing of streaming data. Stream computing differs from traditional computing in that real-time data is processed on the fly by relatively static queries that continuously execute during the lifetime of an application, instead of the data being considered relatively static and all queries being short lived. This is illustrated in Figure 6.4.

In May 2010 IBM released *InfoSphere Streams* or *Streams*. Streams is the result of several year's research conducted by the exploratory stream processing systems group at the IBM T.J. Watson Center. It is a data stream management system middleware designed to ingest, filter, analyze and correlate enormous amounts of data streaming from any number of data sources. Streams is designed to facilitate a rapid response to changing environments leveraging the stream computing paradigm. It has the following objectives [6]:

- Scale using a variety of hardware architectures as demand for processing power changes.
- Provide a platform for handling data streams that is responsive to dynamic user requirements, changing data, and system resource availability.
- Incremental tasking for changing data schemes and types.
- Secure transmission of data streams at all system levels, along with comprehensive auditing of the execution environment.

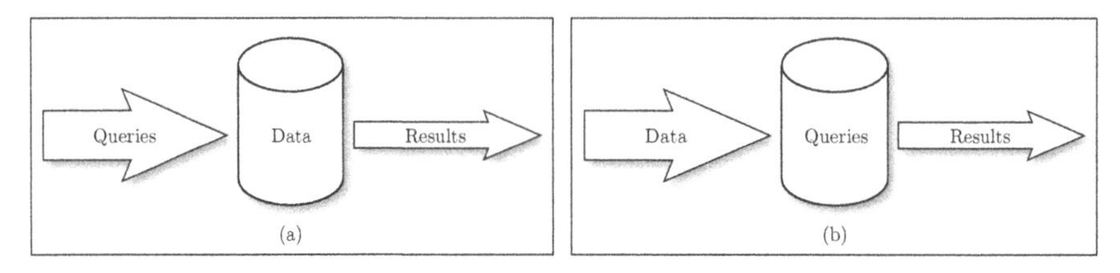

Fig. 6.4 (a) Traditional computing techniques versus (b) stream-computing paradigm [5]

In the remainder of this chapter, we focus on Streams version 1.2[1] which was the platform used in our exploration.

6.2.1 *InfoSphere Streams Terminology and Concepts*

Streams is designed to be highly scalable, so it can be deployed on a single node or on thousands of computing nodes that may have various hardware architectures. The *stream processing core* distributed runtime environment executes numerous long running queries, which Streams refers to as *jobs* [5]. A job can be represented by a *data-flow graph*. Each vertex in the graph represents a *processing element* that transforms the data, and each connecting edge is a data stream, as illustrated in Figure 6.5.

Stream processing elements provide running statistics on their operation. These statistics are utilized by the stream processing core to dynamically optimize job performance by distributing the load and allocating suitable resources for executing each job [7]. Note that a processing element that maps to an underlying computing resource may be changed dynamically by the stream processing core according to load distribution.

The following is a brief description of the stream processing core's main architectural components [8], which is also illustrated in Figure 6.6:

- **Dataflow Graph Manager**
  The *dataflow graph manager* is responsible for the data stream links between the processing elements. Its primary function is to manage the specifications of the input and output ports.
- **Data Fabric**
  The *data fabric* provides the distributed facet of the stream processing. It is made up of a set of daemons that run on each available computing node. The data fabric uses the data stream link specification information from the data-flow graph manager to establish connections between the processing elements and the underlying available computing nodes to transport stream data objects from producer elements to consumer elements.

[1] The current version is 2.0 with similar philosophy but with changes in the programming language

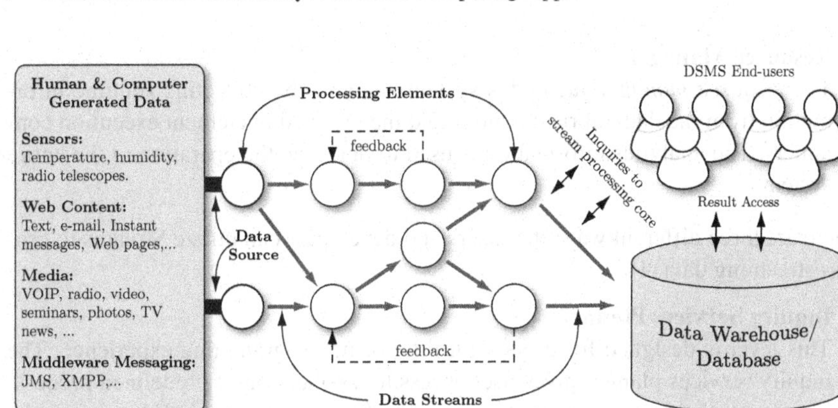

Fig. 6.5 Stream processing core executes numerous long running queries referred to as jobs, which are represented by data-flow graphs [5]

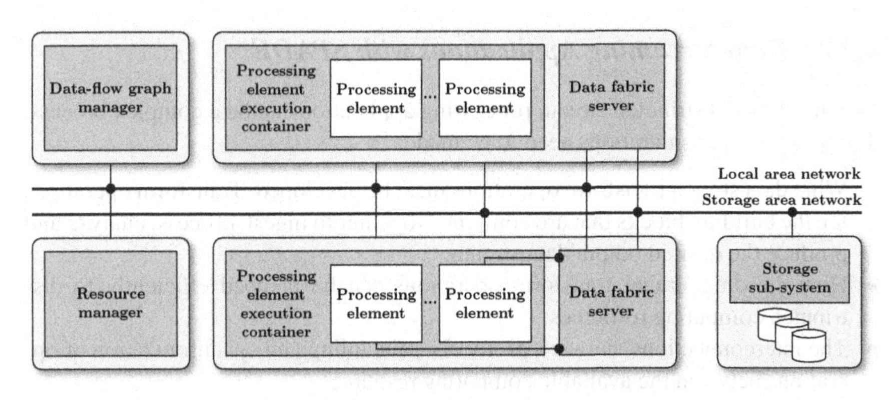

Fig. 6.6 Stream processing core distributed runtime environment [9]

- **Processing Element Execution Container**
  The *processing element execution container* provides the runtime environment and access to the Streams middleware. Furthermore, it also acts as a security fence preventing applications that are running on the processing elements from corrupting the middleware as well as each other.

- **Resource Manager**
  The *resource manager* facilitates system analytics by collecting runtime information from the data fabric daemons and the processing element execution containers. The analytics information is used to optimize the operation of the entire system.

There are three different ways that users and developers can utilize Streams to process streaming data [10]:

- **Inquiry Services Planner**
  This level is designed for users with little or no programming experience. The inquiry services planner gives user access to a collection of predefined processing elements that generate underlying data-flow graphs (behind the scenes the planner generates SPADE applications).
- **Stream Processing Application Declarative Engine (SPADE)**
  SPADE is an intermediate declarative language that enables the construction of data-flow graphs from predefined and custom stream operators.
- **Streams API and the Eclipse Plug-in**
  This is designed for experienced developers who use programming languages such as C++ or Java to implement stream applications that run on the processing elements using the Streams API. Development can be facilitated by using a plug-in available for Eclipse.

6.2.2 Data Streaming Applications with SPADE

Constructing a distributed stream processing application can be a complex process. The following considerations need to be made:

- What data stream transform operations must be developed. Transform operations are the building blocks that are combined together to ingest, process, analyze and produce the desired output data stream.
- How the data stream transform operations can be mapped efficiently to distributed computing resources.
- The interconnections, network protocols, scheduling and synchronization of operations between the available computing resources.

SPADE is designed to deal with these considerations so that programmers can focus on the design of a distributed stream processing application. Using SPADE they can avoid having to develop transform operators as well as face deployment issues that vary depending on the availability of computing resources, network infrastructure and specific technologies [9]. SPADE fulfills its design objectives by collaborating with the stream processing core to provide a dynamic runtime code generation framework capable of achieving scalability and performance through automatic deployment and optimization. This is illustrated in Figure 6.7.

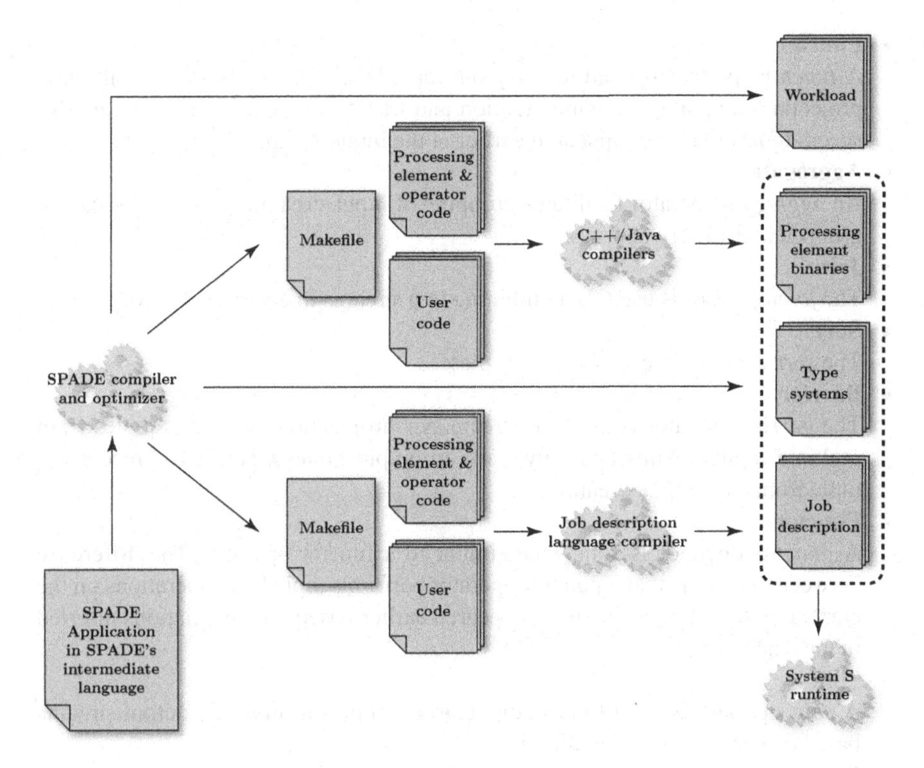

Fig. 6.7 SPADE's code generation framework [9]

The source code of a SPADE application is structured into five main parts:

- **Application Meta Information**
  This part contains the application name and optionally the debug/trace level.
- **Type Definitions**
  In this part the name-spaces and aliases used by the application are declared.
- **External Libraries**
  References to external libraries and header files that contain custom user defined operations are declared in this part. This part is optional.
- **Node Pools**
  In this part pools of computing nodes can be optionally declared. This part is optional since the SPADE compiler can interact with the resource manager to discover available computing node resources.
- **Program Body**
  This is the part where the actual SPADE application is written. In SPADE, streams are considered first class objects where the order of execution is fully characterized by the resulting data streams.

The SPADE language offers the following relational stream operators, used to construct long-running queries:

- **Functor**
  A *functor* operator is used to carry out tuple level operations such as filtering, projection, mapping, attribute creation and transformation. A Functor can also access tuples that have appeared earlier in the input stream.
- **Aggregate**
  An *aggregate* operator facilitates grouping of input stream tuples. Tuples can be grouped in a variety of ways.
- **Join**
  The *join* operator is used for combining two streams in a variety of ways.
- **Sort**
  The *sort* operator is used to order tuples.
- **Barrier**
  The *barrier* operator is used for stream synchronization. It accepts tuples from multiple input streams and only starts to output tuples when it has received a tuple from each input stream.
- **Punctor**
  A *punctor* operator is somewhat similar to a functor operator. The difference between the two is that a punctor operator performs tuple level operations on the current tuple or tuples that have appeared earlier based on punctuations inserted in the data stream.
- **Split**
  A *split* operator is used to pass input stream tuples to multiple output streams based on specified user conditions.
- **Delay**
  The *delay* operator allows a time interval to be specified for delaying a data stream.
- **Edge Adapters**
  Edge adapters are stream operators that function on the boundaries of the SPADE application. They allow a SPADE application to obtain and provide streamed data to applications and entities that are external to the system. There are two types of edge adapter operators:

  – A *source* operator is used to create an incoming data stream of tuples from external data sources.
  – A *sink* operator is used to convert tuples to a format suitable for applications and entities that are external to the system, such as a file system, database, or external application.

- **User Defined Operators (UDOPs)**
  SPADE allows external libraries to be utilized within the SPADE application. Functionality of existing operators can also be extended using UDOPs. UDOPs are developed in C++ or Java using the Streams Eclipse plug-in. UDOPs can be used to port legacy code from other data management platforms into the Streams platform. Furthermore, UDOPs can be used to wrap external libraries from other systems so they can be interfaced with the Streams platform.

- **User Defined Built-in Operators (UBOPs)**
  Although UBOPs allow users to define customized operators they are restricted to the scope of the SPADE application that declares them. On the other hand once defined UBOPs become part of the SPADE language and essentially available for use with any SPADE application.

SPADE also offers advanced features to extend its capabilities and provide a richer platform for data stream application developers.

- **Matrices, Lists and Vectorized Operations**
  Lists and matrices plus the capability to carry out operations on them is a core fundamental feature in many applications such as signal processing, computer graphics, data mining and pattern classification. SPADE offers native support for list and matrix data types as well as vectorized operations which operate on them. Lists or matrices can be created either from external sources via the source operator, functor or punctor operators can be used to create lists or matrices from incoming tuples, or the aggregate operator can create lists or matrices from multiple tuple streams. Many of the SPADE built-in functions are capable of handling matrix, list and scalar type attributes.
- **Flexible Windowing Schemes**
  SPADE supports general windowing mechanisms such as sliding and tumbling windows. SPADE takes these mechanisms further by allowing more sophisticated windowing mechanisms. As an example, an operator can accumulate tuples in a window to hold prior to processing. When a punctuation symbol is received, a processing operation is triggered on tuples currently contained in the window, such as averaging or summing the tuples, and then the window is made to tumble or slide.
- **Per-group Aggregates and Joins**
  Per-group aggregates and joins are designed to cut the number of computations required for operating on a large number of tuple groups. SPADE has the ability to define distinct groupings within a window, so that when a trigger is received an aggregate or join operation can be applied to the entire window or distinct groups within the current window.

6.2.3 Deploying SPADE Applications and Performance Optimization

Discovering the exact optimal mapping (deployment) of a parallelized computer program to loosely coupled (gridded) computing resources is an NP-hard problem [11]. However, heuristics techniques can be used as a practical means for determining an acceptable approximation to an optimal mapping [12]. These heuristics can be improved over time by collecting running statistics that monitor the utilization and performance of computing and network resources.

A SPADE application is a parallelized computer program since it consists of many operators working in parallel towards achieving a common task. InfoSphere Streams approaches optimization of mapping a SPADE program to its underlying computing and network resources in two stages. First, how operators are logically combined (*fused*) into processing elements, and second how processing elements are assigned to physical computing nodes [13].

Info Sphere Streams uses a profiling framework that repeatedly maps processing elements to physical nodes, collects statics and makes necessary remapping adjustments. At the same time the *fusion optimizer* uses the collected statics along with a greedy algorithmic technique to fuse operators into single processing elements.

6.3 Utilizing InfoSphere Streams to Address Large Antennae Array Software Architecture

One of the characteristics of radio astronomy is that it often involves very large volumes of data, particularly when an array of radio telescopes is used for radio interferometry to obtain greater angular resolution of a celestial object. It also has involved many ad-hoc techniques for processing and managing the data.

For instance, the Australian Square Kilometre Array Pathfinder (ASKAP) is a CSIRO-led radio telescope array currently under development at the Murchison radio astronomy observatory. It will consist of 36 antennae, each with a phased array feed that supplies 1.9 Tbps of data and requires 27 Tflops processing to extract a beam visibility. Correlating the resulting 0.6 Tbps data from each antenna is estimated to require 340 Tflops and provide 8 Gbps results for further analysis. ASKAP will have the following the architectural components as given in Fig. 6.8, which illustrates components required to control and manage the data pipeline in a radio telescope array [14]:

Antenna Operations: includes positioning an antenna and setting data acquisition parameters such as sampling rate, bit resolution and filter bank configuration.
Central Processor: correlates the beam visibility data and performs further analysis such as image synthesis or spectral line work.
Array Executive System: responsible for coordinating an observation by the array.
Monitoring Archiver: archives monitoring data generated by the system.
Logging: responsible for logging messages generated by the system.
Data Service: responsible for managing the database storage.
Alarm Management System: manages alarm conditions such as failures.
RFI Mitigation Service: identifies potential sources of radio frequency interference in the received signals.
Ephemerides Service: calculates the positions of celestial objects.
Operator Display: a user interface for system control and monitoring.
Observation Preparation Tool: facilitates the setup and pre-planning of observations.
Observation Scheduler: generates schedules for the execution of observations by the Executive System.

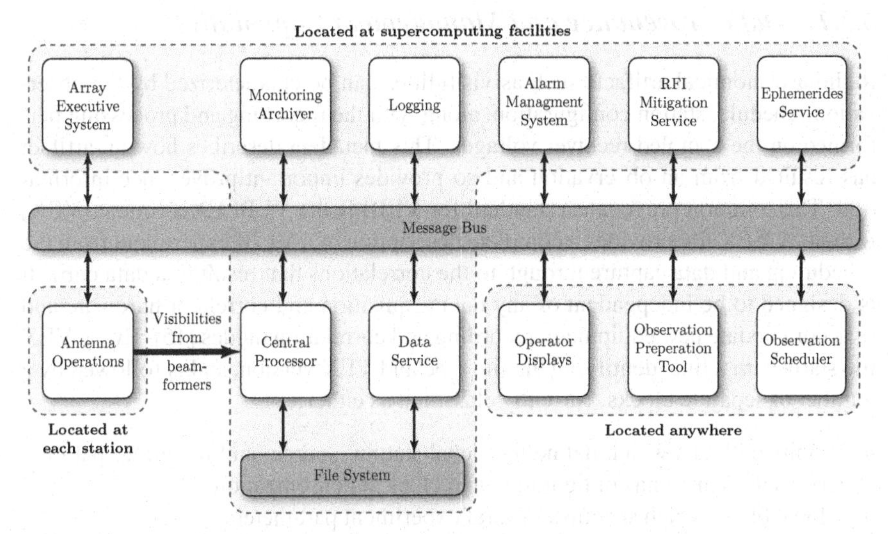

Fig. 6.8 ASKAP top level architectural component view adapted from [14]

Although initially InfoSphere Streams has primarily been applied to the analysis of financial markets, the healthcare sector, manufacturing, and traffic management, it is also suited for managing a radio astronomy pipeline. In particular, the ASKAP architecture in Fig. 6.8 provides a high level abstracted view of how that data pipeline will look like within the ASKAP system. That data pipeline could be instantiated with minimal design effort into streams operators graph, making the transformation of the data in the pipeline transparent. It also allows for the mapping of stream operators to processing elements to be dynamically reconfigurable, important for system scalability, optimizations, and fault tolerance. In fact most of the described architectural components would benefit from these features.

One straightforward scenario for a streams software design mapping the ASKAP architecture is to implement two core main streams instances. The first instance runs at the front-end close to the antennae and is responsible for data conditioning, RFI mitigation, and visibility production. The second instance runs at the central processing unit, and is responsible, among others tasks, for generating images from the visibilities. These two Streams applications could communicate through the Streams middleware services and implement a fast, real-time processing scheme for managing the data from its acquisition all the way through to the analyzed data products.

Other streams jobs could run on the central processing unit. One such job could be responsible for logging, archiving and storage using sink operators. Another could communicate with the scheduling system and send control signals for coordination.

6.3.1 Data Provenance and Management Capabilities

Radio astronomical artifacts such as visibilities can be characterized by the observation schedule, station configuration, along with the recording and processing performed on the sampled receiver voltages. This metadata describes how an artifact has resulted from an observation and so provides important provenance information. The common provenance standard for VLBI is the VLBI EXperiment (VEX) format. A VEX file provides a complete description of a VLBI experiment from the scheduling and data capture through to the correlations that result in a data cube. It is designed to be independent of any data acquisition and correlator hardware and to accommodate new equipment, recording and correlation modes [15]. Every VEX file starts with a line identifying the file type and VEX version, and is followed by a number of separate blocks, currently classified as either:

- Primitive blocks which define low-level station, source, and recording parameters, such as antenna configuration and clock synchronization.
- Global block which specifies general experiment parameters.
- Station and mode blocks which define keywords that combined with the global parameters provide a detailed configuration for an observation at a station.
- Sched block which specifies an ordered list of observations to perform.

For a steady-state radio source the VEX file allows a VLBI experiment to be reproduced, which can be valuable in large arrays where it is impractical to store the huge volumes of raw data. For transient or micro-lensing observations [16, 17] the VEX file provides a basic audit trail for verifying the origins of the experimental results.

However, data provenance practices are less standard for stages after correlation processing. Correlated data can be analyzed in a variety of way, such as for image synthesis or to obtain the power spectral density, and there are not yet standard formats for defining how the resulting data artifacts are produced. The development of a set of provenance standards will be essential for the SKA to ensure the origins of the large number of artifacts produced and for automating their generation.

A radio astronomy system such as ASKAP should be able to provide the following characteristics for managing data:

- Adequate end-to-end throughput not hindered by latency due to the processing elements.
- Intermediate storage capabilities (persistence) to be able to store data summaries in a storage of choice, such as a database or file.
- High availability: the system should be able to work reliably and if failure occurs, data recovery services should be available to avoid important and critical loss of data.

In Streams, adequate throughput can be achieved by a proper architecting of the operators' graph (the graph illustrating how processing elements are linked together to form an application) and by an optimal assignment of processing elements to nodes. This process can only be done on an application-by-application basis. IBM

InfoSphere Streams facilitates the design and optimization of such a graph. Optimizations can be done at compile time, such as those related to the placement of processing elements into nodes or the fusion of processing elements into operators, or at run-time, such as when some nodes get overloaded. Compile-time optimization is efficient when workload and underlying resources are static. Offline profiling of system characteristics prior to deployment is also available, in which case a first pass (prior to deployment) can provide statistics on the data throughput and a second pass (on-deployment) uses those statistics to optimize operators placement and determine when fusion of processing elements should occur.

Data storage can be achieved by use of sink operators capable of storing data to a file, database, or url port. In addition, the software supports user-developed sink operators, useful for custom-based storage needed when sending data to specialized storage recipients.

High availability can be achieved at the middleware level, the application level, and the operator level. At the middleware level, various services are provided to restart a job (potentially on a different node or hardware), replicate name servers across multiple nodes, and monitor activities by writing log files to transactional storage recipients. At the application-level (data processing level), Streams provides checkpointing and automatic restart of processing elements in case of failure. It has tools to provide partial fault tolerance when data loss is a critical issue by means of state persistence capability (the capability to save the state of an operator and restore it).

6.3.2 Some Applications of Streams in Radio Astronomy

There are several examples where Streams has been utilized in the area of radio astronomy and space science.

A space weather monitoring system was developed through joint work by LOIS and IBM Research [18]. It is known that the high-rate, large-volume of near-Earth space data generated by various satellites (such as those of the European Space Agency) is a serious challenge for standard techniques for space weather data monitoring and forecasting. In particular, mining these data in a store-and-process system is not amenable. Streams software was used in [18] to develop a real-time streaming application that measures the intensity, polarization and direction of arrival for signals in the 10 kHz and 100 MHz frequency bands, and on-the-fly generated signal summaries that could be used for space weather forecasting and prevision.

A streaming version of the convolution resampling algorithm was developed by IBM and CSIRO [19] as a prototype imaging application in the Central Processor of ASKAP as described earlier. The version of the algorithm implemented was the w-projection algorithm, which included a CPU intensive gridding step (the process of mapping visibility coordinates into a power of 2 grid). That study showed the flexibility of the streaming software by describing various implementations of streaming scenarios resulting in significant improvements in gridding time.

The next section describes a stream-based autocorrelation approach developed using Streams for data received from a single radio telescope.

6.4 Implementing a Stream-Centric Autocorrelation Data Pipeline & Utilizing Hardware Accelerators

Cross correlation is a fundamental tool in radio astronomy since it helps with identifying repeating patterns obscured by the predominant random noise content of extraterrestrial signals. As these signals are mostly composed of random noise they can be characterized as stationary stochastic processes where the mean and variance do not change over time.

An autocorrelation is the cross correlation of a signal with itself. Autocorrelation is mainly used for single antenna applications and calibration of individual arrayed antennae. Implementing autocorrelation requires less effort since only one signal is considered and therefore no time delays are required. As a consequence of its relative simplicity implementing an autocorrelation pipeline is a logical starting point for constructing a basic cross correlation pipeline for radio astronomy signal processing.

6.4.1 Autocorrelation and the Power Spectral Density in Radio Astronomy

If voltage samples $s(t)$ are obtained from an antenna the energy spectral density $E(f)$ of the incident electromagnetic waves can be determined. The energy spectral density is the energy carried by the incident waves per unit frequency f, which is given by the Fourier transform:

$$E(f) = \left| \int_{-\infty}^{\infty} s(t) e^{-2\pi i f t} \, dt \right|^2 .$$

However, $s(t)$ is a stationary signal and is not square integrable so its Fourier transform does not exist. Instead the *Wiener-Khinchin* theorem is applied to obtain the power spectral density (PSD) of the voltage signal from the autocorrelation function $r(\tau)$:

$$r(\tau) = < s(t) s(t-\tau) > .$$

The Wiener-Khinchin theorem states that the PSD $P(f)$ of the signal $s(t)$ is the Fourier transform of the autocorrelation function $r(\tau)$:

$$P(f) = \int_{-\infty}^{\infty} r(\tau) e^{-2\pi i f t} \, dt.$$

The PSD $P(f)$ is the power carried by the incident waves per unit frequency f.

From Figure 6.9 the PSD can be obtained by either performing an FX or XF correlation. An FX correlation is a Fourier transform followed by element-wise

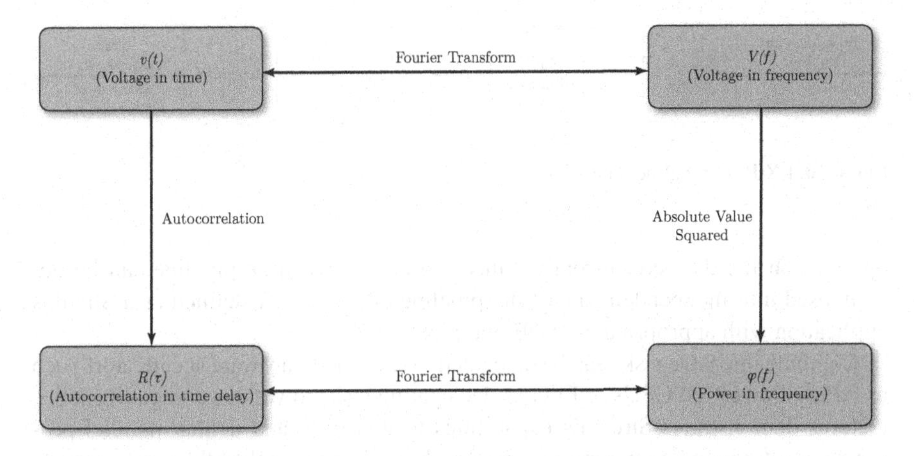

Fig. 6.9 Relation between voltage in time and frequency domains with the autocorrelation function and power spectral density [20]

multiplication. An XF correlation is a cross multiplication followed by a Fourier transform. FX style correlation is preferred for software implementations since it involves fewer multiplications [21].

6.4.2 Implementing a PSD Pipeline as a Stream Based Application

Analogue voltage signals on the antenna receiver are sampled and digitized by an analogue to digital converter. The digitized real value data (2-16 bit digitization) are then streamed in real-time into an FX style pipeline to produce the power spectral density (PSD) of the signal. The FX PSD pipeline illustrated in Figure 6.10 is comprised of the following steps:

- Collect digitized signal data into chunks whose size is determined by the amount of data optimally processed together in the pipeline.
- *Channelize* each chunk to obtain frequency domain data by applying a Fast Fourier Transform to obtain single-precision float complex value data chunks.
- Obtain the autocorrelation of the data in the frequency domain by multiplying each complex value in a chunk by its complex conjugate.
- Integrate and average the data chunks over time to obtain a best PSD estimate of the signal.

The final stage plays an important role in improving the signal to noise ratio, hopefully allowing buried coherent signals of interest to emerge from the predominantly random noise polluted signal.

The entire pipeline can be viewed as a parallel program where each stage of the pipeline is an independent task. As data flow through the pipeline each stage

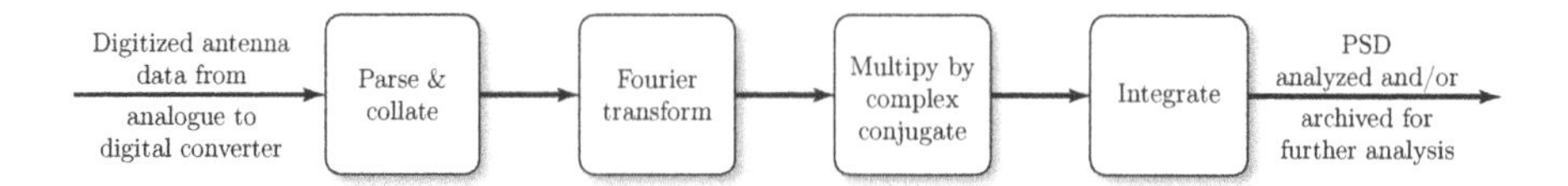

Fig. 6.10 FX PSD pipeline data flow

operates on the data concurrently. Since the auto-correlation pipeline can be decomposed into independent stages the pipeline can be easily defined as a Streams application with appropriate SPADE operators.

Compute intensive tasks can be delegated to specialist hardware accelerators such as GPUs, PowerCell CPUs or FPGAs. Delegating tasks to various computing architectures demonstrates Stream's capabilities to construct and deploy parallel programs to heterogeneous computing clusters. For the PSD pipeline the most compute intensive task is the Fourier transform, and will be assigned to a hardware accelerator for processing.

The SPADE application makes use of *virtual streams* for predefining the various tuple data structures, known as a tuple's *schema*, that are utilized by its underlying stream operators. Virtual streams contribute towards ease of programming as well as understandability of the tuple structure flowing between stream operators. The PSD SPADE application declares the following virtual streams:

`RawData(data:ShortList)` defines the schema used for creating tuples arising from ingesting and parsing real value integer radio astronomy antenna data. The bit resolution used to digitize the antenna analogue signal may vary from as low as 2 bits up to 16 bits according to the recording system and type of observation. Essentially, using a `ShortList` data type satisfies the bit-level representation requirements for most radio astronomy recording formats.

`RawDataChunk(acceleratorID:Integer, schemaFor(RawData))` defines the structure for a data chunk designated to a specific accelerator server for channelization.

`ChannelData(real:FloatList, imag:FloatList)` defines the structure for channelized data chunks that have been channelized by an accelerator server.

`PowerSpectrumData(psd:FloatList)` defines the structure for tuples containing PSD data. This virtual stream's schema is used by several operators for producing the different integration stages of the PSD.

Altogether the autocorrelation spectrometer SPADE application uses seven distinct stream operators. The number of actual operators depends on the number of accelerators utilized for channelization and integration stages. Figure 6.11 shows the data flow graph for the SPADE PSD application (see Appendix for SPADE code listing).

The first stream operator in the application is a Source operator which ingests digitized unsigned integers from the signal. The data are parsed by this operator according to the format used to digitize and pack samples. The Source operator

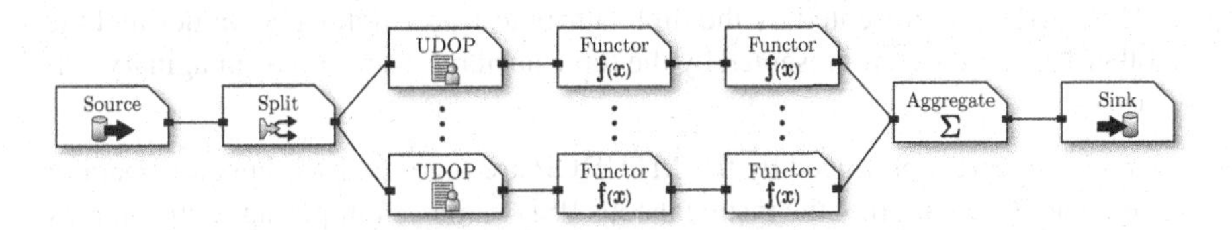

Fig. 6.11 FX PSD pipeline data flow graph and SPADE operators

builds a tuple according to the `RawDataChunk` virtual stream schema, reading into the `ShortList` data tuple attribute. Each `RawDataChunk` tuple is assigned to an accelerator server by assigning a positive integer to the `acceleratorID` tuple attribute. The number of tokens contained in the data attribute is given by the size and number of FFTs to be performed by the channelization stage. A positive integer value between 0 and the total number of accelerators utilized is assigned to the `acceleratorID` attribute in a round robin fashion.

Tuples resulting from a Source operator are ingested by a Split operator. Essentially the Split operator is a multiplexed stream. Each sub-stream in the multiplex stream carries `RawDataChunk` tuples according to their respective identifier. Effectively the Split enables `RawDataChunk` tuples to be fanned out to several accelerators.

The Streams Processing Core may not be supported on a particular accelerator architecture, so one way around is to use a UDOP. To enable asynchronous communication with the accelerator server multi-threaded UDOPs (MTUDOP) are adopted by the SPADE application. An MTUDOP facilitates uncoupling the processes of receiving and transmitting tuples. For further versatility the UDOP uses configuration switches so that the same UDOP can be reused.

The MTUDOP uses three switches allowing the SPADE application to configure its operation with respect to which accelerator will be used for processing as well as the communication mode for incoming and outgoing data. During the initialization phase the MTUDOP extracts configuration information from the switch operators and establishes incoming and outgoing connections.

Once the connections have been made to a specific accelerator the MTUDOP runs two processes:

- The input tuple process ingests tuples transmitted by a specific Split operator sub-stream. The ingested tuples are converted to floats since software implementations for Fourier transforms require this. Following that some byte reordering may be necessary depending on the architecture of the accelerator. Once the type conversion and byte reordering are accomplished the data are sent to the accelerator for channelization.
- The output tuple process receives data from the accelerator. Similarly to the previous process, received data may need byte reordering. The `ChannelData` schema is used to define the outgoing tuple structure. The channelized data chunk received from the accelerator arrives in interleaved complex number format, and so real and imaginary parts are separated into two `FloatList` data types. One

FloatList represented by the tuple attribute real holds real values and the other FloatList represented by the tuple attribute imag holds imaginary values.

ChannelData produced by the MTUDOP are ingested by a Functor operator responsible for computing the instantaneous PSD values, multiplying each complex number by its complex conjugate to produce a real value. The resulting real values are defined by the PowerSpectrumData virtual stream schema.

Each PowerSpectrumData tuple arising from the first Functor operator contains data for several FFT problems, so the second Functor operator integrates those results. Effectively the second Functor integrates multiple FFT problems contained in a single data chunk. To accomplish integration within a data chunk the second Functor operator uses customized user-defined logic. Integrations within a data chunk are performed using a *Slice* operation that helps with extracting the result of each FFT problem for summing and averaging. Summing and averaging produces a FloatList and so the same PowerSpectrumData schema is used to represent the resultant tuples.

PowerSpectrumData tuples from the second Functor operator are then integrated with an Aggregate operator. The aggregation count is specified by the SPADE application. Fundamentally the aggregation count is the required integration time. The longer the integration time, that is the higher the aggregation count, the better the signal to noise ratio.

Integrated power spectral density tuples produced by the Aggregate operator and defined by the PowerSpectrumData schema are ingested by a Sink operator. The Sink operator may either write the integrated PSD strips to disk or possibly stream them over the network for deeper analysis or visualization.

6.4.3 Using Accelerators (Heterogeneous Computing)

An accelerator is intended to provide specialized accelerated computing services to assist with handling compute intensive operations. The objective for utilizing accelerators is to enable real-time data management operations especially for areas that involve processing large amounts of data. An important consideration that must be made when using a particular accelerator hardware is the use of its unique performance primitives and libraries. Neglecting this consideration in many cases will lead to ineffective utilization of the accelerator's intensive computing capabilities.

The SPADE PSD application described previously is designed to function with any type of accelerator. At the time of implementation the PowerCell CPU accelerator was available. In this subsection we describe how the compute intensive Fourier transform was implemented on the PowerCell CPU using its unique performance primitives and libraries.

The Fourier transform service is provided by an implementation of the discrete Fast Fourier Transform (FFT) on a PowerCell QS22 Blade Server. The QS22 Blade Server comprises of two PowerCell CPUs. Each PowerCell CPU comprises of nine cores; one 64 bit duo-core PowerPC processor and eight 128 bit RISC processors.

The FFT server program is a multi-threaded application that executes the following four threads:

- Applications main thread; responsible for initializing the FFT memory buffers working area as well as starting the receiving, processing and sending threads. After initialization the main thread blocks until the application is terminated.
- The receiving thread creates a server socket and listens for a single client connection. Once a sender client connects, the FFT Server receives data chunks containing multiple FFT blocks. Each data chunk is written to a specific buffer, which is then flagged to indicate that it is ready for processing.
- The processing thread performs a real to complex FFT on memory buffers that have been flagged as fully received by the receiver thread. Consequently since the FFT is real to complex then ultimately the same amount of data received will be the same amount sent. The FFT is accomplished using all 16 SPU cores on a QS22, hence the reason why a single data chunk contains multiple FFT blocks. Once a single data chunk contained in a given buffer has been processed it is then flagged by the processor thread to indicate that the results are ready for sending.
- The sending thread sends the contents of memory buffers that have been flagged as processed. Memory buffers that have been sent are flagged to indicate that the buffer can be reused for receiving.

The application makes use of a multi-buffering scheme for allowing the receiving, processing and sending threads to operate in an asynchronous fashion. The threads operate asynchronously as long as there are buffers available. Any contribution towards asynchronous operation between concurrent threads reduces blocking thereby contributing to a gain in overall performance. Nevertheless mitigating concurrency between thread access to each individual buffer is still required, and this is accomplished via a two-phase locking mechanism. The processing thread makes use of the SDK for Multicore Acceleration FFT library to efficiently compute a large number of FFT problems in parallel. The FFT library achieves significant computational performance gains by exploiting the PowerCell CPU's vectorized SIMD capabilities utilizing two main approaches:

- **Striping across vector registers**
  The SPE's architecture is 128-bit hence its underlying Synergistic Processing Unit (SPU) registers can be considered as vector registers. In the case of this implementation the FFTs are performed using 32 bit (single-precision) floating point values. A single SPU register can therefore hold four individual 32-bit floats and operate on the entire vector of floats using a single instruction. When performing FFT operations rather than loading four values from one FFT problem, four values are loaded from four problems. This technique is known as *striping* multiple problems across a single register. Striping values from four problems across a register enables these problems to be accomplished in unison, and the code used for indexing and twiddle calculations can be reused [22].

  The FFT algorithm makes extensive use of compute intensive trigonometric mathematical operations. Fortunately the SIMD Math Library contains vectorized versions of common mathematical operations. Utilizing vectorized math operations along with data stripping across vector registers dramatically reduces the frequency of their usage. However data striping values across vector registers is limited since it requires all the values from the FFT problems to be present in the SPU's LS, which has maximum capacity of only 256 kB. Hence data striping across vectors is limited to small point size FFT problems.

- **Vector synthesis**
  To accommodate the memory bound limitations imposed by a SPU's LS, SPU *shuffle* operations can rearrange a large set of FFT problems residing in main memory into vector form prior to DMA transfer to the SPUs. This rearrangement of scalar FFT problem values into vector form is known as *vector synthesis*, and is illustrated in Figure 6.12. Subsequently, once the SPUs complete FFT computations of the large set of problems, the vectorized values must be rearranged back into scalar form. Naturally the rearrangement of a large data set of FFT problems into vectors then back to scalars does incur a computational expense. However since the large data set is prepared for vectorized trigonometric operations then the gains made in reducing the amount of computational intensive trigonometric operations greatly outweigh the costs incurred by vector synthesis operations [22, 23].

To gain further significant performance speed-ups (of almost 10 times) on the Cell B.E. architecture, the implementation uses the huge translation lookaside buffer (TLB) file system. Huge TLB page files allocate large pages (16 MB per page) of contiguous memory. Utilizing huge page files reduces the TLB miss rate and consequently leads to a gain in performance. The data chunk size mentioned in Section 6.4.2 was set to 16 MB to fully utilize an entire huge TLB page.

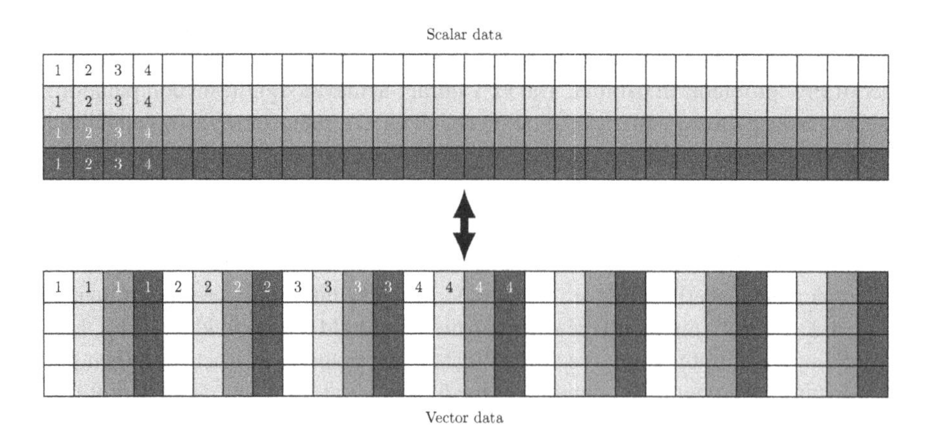

Fig. 6.12 Scalar versus vector data arrangement in a contiguous block of main memory [22]

6.4.4 *Testing the SPADE PSD Application*

The SPADE PSD application was tested using network streamed data from the AUT University 12m radio telescope located at Warkworth in New Zealand. An IBM Blade Center holding x86 HS12 and dual-PowerCell QS22 blades was used to run the application. The HS12 blade was used to execute the SPADE PSD application. As the PowerCell FFT library is limited to a maximum size of 8192 points (for real to imaginary number FFT transforms) each huge TLB page could accommodate 512 FFT problems.

Test data were sampled by the radio telescope from the European Space Agency Mars Express Orbiter at 60 MHz (32 MHz bandwidth due to Nyquist criterion) using 8-bit digitization. The resulting power spectral density is illustrated in Figure 6.13.

The overall shape of the distribution is dictated by the specific telescope and its receiving system. This shape was also determined independently using a hardware spectrum analyzer to verify the correctness of the PSD application. Of particular interest in the spectrum was the 8420.4321 MHz signal detected by the application which was being emitted by the orbiter at the time of the test.

On average the application took 8 ms to autocorrelate 512 FFT problems of size 8192 with a just single accelerator, a performance of 30 Gflops. Although greater performance could also be achieved by either using an FFT implementation tuned for the specific size or by utilizing more accelerators, the application could already handle 524 MHz sampling from a single antenna. However, careful attention must be paid to the networking layers in order to utilize the potential power of any accelerator. During initial testing the QS22 blade accelerators utilized TCP over full duplex Gigabit Ethernet links and both the average data chunk inter-arrival and inter-departure times were found to be 0.6 s, limiting the sampling rate to 6 MHz per link.

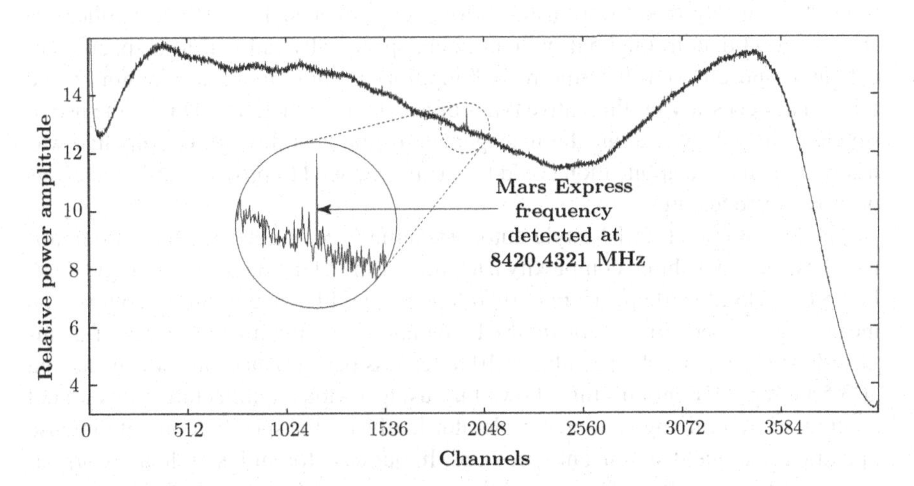

Fig. 6.13 Plot of the PSD results obtained from a 40 second observation of the ESA Mars Express spacecraft conducted by the AUT 12m radio telescope

This limitation can be removed by either utilizing multiple links in a round robin fashion from an x86 blade, utilizing an alternative protocol to TCP, or employing an alternative networking technology such as Infiniband.

6.4.5 Performance and Scalability

The results from the previous subsection demonstrate that Streams shows good performance calculating the PSD using a mix of HS12 and QS22 blades with the PowerCell as the accelerator. In particular, the greatest performance limitation was determined to be inter-blade networking rather than anything associated with the Streams framework itself, despite Streams being hardware and network technology agnostic. Streams allows the underlying computer and network technologies to be changed for best suiting the computations required in a particular application. This enables it to leverage the performance of new hardware as it becomes available, while reducing the effort to reengineer software applications.

The scalability features of Streams are valuable for meeting growing computational demands. The SPADE PSD application can scale to handle greater sampling rates, higher frequency resolution via a larger FFT point size, or additional antennae. The SPC performs dynamic assignment of processing elements to physical nodes which enables the SPADE application to dynamically meet the demands of intensifying computations. This ability to dynamically redeploy a parallel program during runtime to physical nodes allows Streams to scale effectively.

6.5 Conclusion

This application successfully demonstrated the viability of implementing a real-time PSD entirely in software using InfoSphere Streams. The SPADE application showed good data throughput without being specifically tailored to a specific accelerator, and allowed dynamic reconfiguration to allow more accelerators to be utilized as necessary or alternative types of accelerators included. Due to the use of standard SPADE operations the management of the data through the pipeline was transparent and the application could be easily extended to provide further analysis or provenance features.

The operations of an FX style autocorrelating spectrometer pipeline are dominated by the algorithmic complexity $n\log_2(n)$ of the FFT operation. A single dual-PowerCell QS22 blade measured 30 GFlops for the FFT operation. In comparison measuring the performance of the FFT operation utilizing all four cores of a single x86 HS12 blade achieved almost 10 times less performance, and was measured at 3.5 GFlops. Essentially this shows that using various architectures for parallel computing by utilizing suitable accelerator hardware for specific compute intense operations can yield significant speed ups. In our case for an FX style autocorrelation pipeline a speed up of almost 10 times was achieved per QS22 blade.

IBM Info Streams proved it's flexibility to operate using various architectures in unison. Despite the I/O bound links Streams was capable of maximizing the link bandwidth as well as manage the data flow without information loss. Implementing the PSD pipeline in SPADE allowed parameters such as the integration time and FFT point size to be changed in real-time without compromising the flow of data.

Streams facilitates both implicit and explicit parallelization. Implicit parallelization is achieved by fusing operators into processing elements, and explicit parallelization by deployment of processing elements to many physical nodes. Furthermore Streams goes beyond conventional parallelization middleware and frameworks such as MPI (Message Passage Interface) and OpenMP by allowing dynamic operator fusing and processing element deployment to physical nodes during runtime. This degree of dynamic operation enables Streams to provide on demand scalability to increasing data loads and computations.

In this work we mainly focus on reviewing InfoSphere Streams and its potential use for Radio Astronomy. In our opinion the Streams approach has shown positive results to warrant further research and serious consideration for managing DPDM aspects of large antennae arrays. Further research is required to conduct more formal and specific comparative analysis between Streams and other middleware such as MPI and ICE (Internet Connection Engine). Another interesting area that requires more rigorous investigation is Streams scaling capabilities using a larger cluster of x86 nodes as well as combining other accelerators such as GPUs, Intel MIC (Many Integrated Core) architecture and FPGAs.

Acknowledgements. The first author would like to thank the New Zealand government's Tertiary Education Commission Build IT fund for funding this research and the IBM cooperation for providing the computing hardware under the shared university research grant.

Appendix

The following SPADE code listing shows an implementation of an autocorrelation spectrometer application. Lines 4-7 define the virtual streams used by the applications stream operators. In lines 8-11 is a user defined source operator responsible for receiving network digitized raw antenna data using TCP. In lines 12-15 a split operator is used to distribute data to the PowerCell accelerators (for simplicity this listing uses two accelerators). In lines 17-20 an MTUDOP is used to send time series data and receive frequency domain data to and from the PowerCell accelerators. The frequency domain data received from a particular accelerator is then autocorrelated by a functor operator in lines 21-24. Since the accelerators are given a chunk containing multiple FFTs lines 25-41 integrate this chunk. In lines 43-46 an aggregate operator is used to integrate averaged chunks. In this particular listing the aggregate operator sends a result to the sink operator in line 47 every time it integrates 523 averaged PSD chunks.

```
1  [Application]
2   AutoCorrelator
3  [Program]
4   vstream RawData(data : ShortList)
5   vstream RawDataChunk(cellID: Integer, schemaFor(RawData))
6   vstream ChannelData(real : FloatList, imag : FloatList)
7   vstream PowerSpectrumData(psd : FloatList)
8   stream Antenna(schemaFor(RawDataChunk))
9    := Source() ["stcp://thishost:9932/",
10   udfBinFormat="AntennaParser",
11   blockSize=8*1024] {}
12  for_begin @Blade_ID 1 to 2
13    stream QS22@Blade_ID(schemaFor(Antenna))
14  for_end
15   := Split(Antenna)[cellID]{}
16  for_begin @Blade_ID 1 to 2
17    stream FFT@Blade_ID(schemaFor(ChannelData)) :=
18       MTUdop(QS22@Blade_ID)["MT_QS22_FFT"] {
19           switch1="@Blade_ID", switch2="9933", switch
                3="9934"
20        }
21    stream PSD@Blade_ID(schemaFor(PowerSpectrumData))
22       := Functor(FFT@Blade_ID) [] {
23           psd := apply(pow, real, 2.0) .+ apply(pow, imag,
                 2.0)
24       }
25    stream IntegrateChunk@Blade_ID(schemaFor(
26       PowerSpectrumData)) := Functor(PSD@Blade_ID)
27    <
28       Integer $count := 1;
29       FloatList $average := makeFloatList();
30    >
31    <
32      $average := slice(psd, 0, 4096);
33      while($count < 512) {
34        $average := $average .+ slice(psd, $count * 4096,
             4096);
35        $count := $count + 1;
36       }
37       $average := $average ./ 512.0;
38       $count := 1;
39    >
40    [true]
41    {psd := $average}
42  for_end
43  stream Integrate(schemaFor(PowerSpectrumData)) :=
44   Aggregate(IntegrateChunk1, IntegrateChunk2 <count(523)>)
         [] {
```

```
45        psd := Avg(psd)
46    }
47  Nil := Sink(Integrate)["file:///../data/spectrum.bin",
        nodelays, udfBinFormat="DataFormatter"] {}
```

References

1. Walker, R.: What the VLBA Can Do for You: Capabilities, Sensitivity, Resolution, and Image Quality. In: Zensus, J.A., Diamond, P.J., Napier, P.J. (eds.) Very Long Baseline Interferometry and the VLBA. ASP Conference Series, vol. 82, pp. 133–157 (1995)
2. Weston, S.: Development of Very Long Baseline Interferometry (VLBI) techniques in New Zealand: Array simulation, image synthesis and analysis. M.Phil thesis, Auckland University of Technology (2008), http://hdl.handle.net/10292/449
3. van der Schaaf, K., Broekema, C., Diepen, G., Meijeren, E.: The LOFAR central processing facility architecture. Experimental Astronomy 17(1-3), 43–58 (2004)
4. Varbanescu, A.L., van Amesfoort, A., Cornwell, T., van Diepen, G., van Nieuwpoort, R., Elmegreen, B., Sips, H.: Building high-resolution sky images using the Cell/B.E. Scientific Programming 17, 113–134 (2009)
5. IBM Corporation. System Infrastructure for Streaming (2009), Retrieved from http://domino.research.ibm.com/comm/research_projects.nsf/pages/esps.Projects.html
6. Bollard, C., Farrell, D.M., Lee, M., Stone, P.D., Thibault, S., Tucker, S.: IBM InfoSphere Streams: Harnessing Data in Motion. IBM Redbooks: International Business Machines Corporation (2010), http://www.redbooks.ibm.com/redbooks/pdfs/sg247865.pdf
7. Turaga, D.S., Verscheure, O., Chaddhari, U.V., Amini, L.D.: Resource Management for Networked Classifiers in Distributed Stream Mining Systems. In: Proceedings of the Sixth International Conference on Data Mining, pp. 1102–1107 (2006), doi:10.1109/ICDM.2006.136
8. Jain, N., Amini, L., Andrade, H., King, R., Park, Y., Selo, P., Venkatramani, C.: Design, implementation, and evaluation of the linear road benchmark on the stream processing core. In: Proceedings of the ACM SIGMOD International Conference on Management of Data (2007), doi:10.1145/1142473.1142522
9. Gedik, B., Andrade, H., Wu, K., Yu, P.S., Doo, M.: SPADE: the system S declarative stream processing engine. In: Proceedings of the ACM SIGMOD International Conference on Management of Data (2008), doi:10.1145/1376616.1376729
10. IBM Corporation. System S - Stream Computing at IBM Research (2009), Retrieved from http://public.dhe.ibm.com/software/data/sw-library/ii/whitepaper/SystemS_2008-1001.pdf
11. Andersson, J., Ericsson, M., Löwe: An Adaptive High-Performance Service Architecture. Software Technology Group, MSI. Växjö universitet (2003)
12. Culler, D., Karp, R., Patterson, D., Sahay, A., Schauser, K.E., Santos, E., Subramonian, R., von Eicken, T.: LogP: Towards a Realistic Model of Parallel Computation. In: 4th ACM PPOPP,5/93/CA, USA (1993)
13. Gedik, B., Andrade, H., Wu, K.: A Code Generation Approach to Optimizing Distributed Data Stream Processing. In: ACM CIKM 2009, Hong Kong, China, Novemebr 2-6 (2009)

14. Guzman, J.C., Humphreys, B.: The Australian SKA Pathfinder (ASKAP) Software Architecture. In: Proceedings of SPIE, vol. 7740, p. 77401J (2010)
15. VEX File Definition. VLBI Standards & Resources Website, Retrieved from `http://vlbi.org/vex/`
16. McLaughlin, M.: Rotating Radio Transients. In: Becker, W. (ed.) Neutron Stars and Pulsars, pp. 41–66. Springer, Berlin (2009)
17. Refsdal, S.: The gravitational lens effect. Monthly Notices of the Royal Astronomical Society 128, 295 (1964)
18. Daldorff, L.K.S., Mohammadi, S.M., Bergman, J.E.S., Thide, B., Biem, A., Elmegreen, B., Turaga, D.S.: Novel data stream techniques for real time HF radio weather statistics and forecasting. In: Proceedings of IRTS, Edinburgh, UK, April 28-30 (2009) ISBN: 978 1 84919 123 4
19. Biem, A., Elmegreen, B., Verscheure, O., Turaga, D., Andrade, H., Cornwell, T.: A streaming approach to radio astronomy imaging. In: Proceedings of IEEE ICASSP, pp. 1654–1657 (2010), doi:10.1109/ICASSP.2010.5495521
20. Rohlfs, K., Wilson, T.L.: Tools of Radio Astronomy, 4th edn., pp. 50–52. Springer, Heidelberg
21. Bunton, J.D.: New Generation Correlators. In: Proceedings of the XXVIIth General Assembly of International Union Radio Science (URSI), Commission J06, Vigyan Bhavan, New Delhi, India, October 23-29 (2005)
22. Arevalo, A., Matinata, R.M., Pandian, M., Peri, E., Ruby, K., Thomas, F., Almond, C.: Programming the Cell Broad Engine Architecture: Examples and Best Practices, 1st edn. IBM Redbooks: International Business Machines Corporation (2008)
23. Lu, J., Nobels, A., Perrone, M.: IBM Research Report: Accelerating FFT Performance Using the Cell BE Processor. T. J. Watson Research Center, Yorktown Heights. IBM Research Division, New York (2007)

Chapter 7
Using Provenance to Support Good Laboratory Practice in Grid Environments

Miriam Ney, Guy K. Kloss, and Andreas Schreiber

Abstract. Conducting experiments and documenting results is daily business of scientists. Good and traceable documentation enables other scientists to confirm procedures and results for increased credibility. Documentation and scientific conduct are regulated and termed as "good laboratory practice." Laboratory notebooks are used to record each step in conducting an experiment and processing data. Originally, these notebooks were paper based. Due to computerised research systems, acquired data became more elaborate, thus increasing the need for electronic notebooks with data storage, computational features and reliable electronic documentation. As a new approach to this, a scientific data management system (DataFinder) is enhanced with features for traceable documentation. Provenance recording is used to meet requirements of traceability, and this information can later be queried for further analysis. DataFinder has further important features for scientific documentation: It employs a heterogeneous and distributed data storage concept. This enables access to different types of data storage systems (e. g. Grid data infrastructure, file servers). In this chapter we describe a number of building blocks that are available or close to finished development. These components are intended for assembling an electronic laboratory notebook for use in Grid environments, while retaining maximal flexibility on usage scenarios as well as maximal compatibility overlap towards each other. Through the usage of such a system, provenance can successfully be used to trace the scientific workflow of preparation, execution, evaluation, interpretation and archiving of research data. The reliability of research results increases and the research process remains transparent to remote research partners.

Miriam Ney · Andreas Schreiber
Simulation and Software Technology, German Aerospace Centre,
Berlin, Cologne, Germany
e-mail: NeyMiriam@googlemail.com, Andreas.Schreiber@dlr.de

Guy K. Kloss
School of Computing + Mathematical Sciences, Auckland University of Technology,
Auckland, New Zealand
e-mail: Guy.Kloss@aut.ac.nz

Q. Liu et al. (Eds.): Data Provenance and Data Management in eScience, SCI 426, pp. 157–180.
springerlink.com

7.1 Introduction

With the "Principles of Good Laboratory Practice and Compliance Monitoring" the OECD provides research institutes with guidelines and a framework to ensure good and reliable research. It defines "Good Laboratory Practice" as *"a quality system concerned with the organisational process and the conditions under which non-clinical health and environmental safety studies are planned, performed, monitored, recorded, archived and reported"* (p. 14 in [8]). This definition can be extended to other fields of research. To prove the quality of research is of relevance for credibility and reliability in the research community. Next to organisational processes and environmental guidelines, part of the good laboratory practice is to maintain a laboratory notebook when conducting experiments.

The scientist documents each step, either taken in the experiment or afterwards when processing data. Due to computerised research systems, acquired data increases in volume and becomes more elaborate. This increases the need to migrate from originally paper-based to electronic notebooks with data storage, computational features and reliable electronic documentation. For these purposes suitable data management systems for scientific data are available.

7.1.1 A Sample Use Case

As an example use case a group of biologists are conducting research. This task includes the collection of specimen samples in the field. Such samples may need to be archived physically. The information on these samples must be present within the laboratory system to refer to it from further related entries. Information regarding these samples possibly includes the archival location, information on name, type, date of sampling, etc.

The samples form the basis for further studies in the biological (wet) laboratories. Researchers in these environments are commonly not computer scientists, but biologists who just "want to get their research done." An electronic laboratory notebook application therefore must be similarly easy to operate in day-to-day practice like a paper-based notebook. All notes regarding experimentation on the samples and further derivative stages (processing, treatments, etc.) must be recorded, and linked to a number of other artifacts (other specimen, laboratory equipment, substances, etc.).

As a result of this experimentation further artifacts are derived, which need to be managed. These could be either further physical samples, or information (data, measurements, digital images, instrument readings, etc.). Along with these artifacts the team manages documents outlining the project plan, documents on experimental procedures, etc.

In the end every managed artifact (physical or data) must be linked through a contiguous, unbroken chain of records, the provenance trail. The biologists in our sample use case cooperate with researchers from different institutes in different (geographical) locations. Therefore, the management of all data as well as provenance must be enabled in distributed environments, physically linked through the Internet.

The teams rely on a common Grid-based authentication, which is used to authorise principals (users, equipment, services) across organisational boundaries.

The recorded provenance of all managed artifacts can be used in a variety of ways. Firstly, it is useful to document and *prove* proper scientific procedures and conduct. Beyond this compliance requirement provenance information can be used in further ways: It enables often previously not possible (or very tedious) ways of analysis. By querying the present provenance information, questions can be answered which depend on the recorded information. These questions may include some of the following:

- *Question for origin:* What artifacts were used in the generation of another artifact?
- *Question for inheritance:* What artifacts and information were generated using a given artifact?
- *Question for participants:* What actors (people, devices, applications, versions of tools, etc.) were employed in the generation of an artifact?
- *Question for dependencies:* Which resources from other projects/processes have been used in the generation of an artifact?
- *Question for progress:* In what stage of a processing chain is a given artifact? Has the process the artifact is part of been finalised?
- *Question for quality:* Did the process the artifact is part of reach a satisfactory conclusion by some given regulations or criteria?

7.1.2 Data Management with the DataFinder

In order find a solution to common data management problems, the German Aerospace Centre (DLR) – as Germany's largest research institute – developed an open source data management application aimed at researchers and engineers: *DataFinder* [17, 20]. DataFinder is a distributed data management system. It allows heterogeneous storage back-ends, meta-data management, flexible extensions to the user interface and script-based automation. To implement required features for reliable and auditable electronic documentation provenance technologies can be used [5].

When analysing the data management situation in scientific or research labs, several problems are noticeable:

- Each scientist individually is solely responsible for the data generated and managing it as deemed fit. Often others cannot access it, and duplication of effort may occur.
- If a scientist leaves the organisation, it is possible that no one understands the structure of the data left behind. Information can be lost.
- Researchers often spend a lot of time searching for data. This waste of time decreases productivity.
- Due to long archiving periods and an increasing data production rate, the data volume to store increases significantly.

To overcome this situation common in many research institutes, the DLR facility Simulation and Software Technology has developed the scientific data management system DataFinder (cf. [20]).

7.1.2.1 General Concepts

DataFinder is an open source software written in Python. It uses a server and a client component. The server component holds data and associated meta-data. Data and meta-data is aggregated in a shared data repository and accessed and managed through the client application. Fig. 7.1 shows the user interface of the DataFinder, when connected to a shared repository.

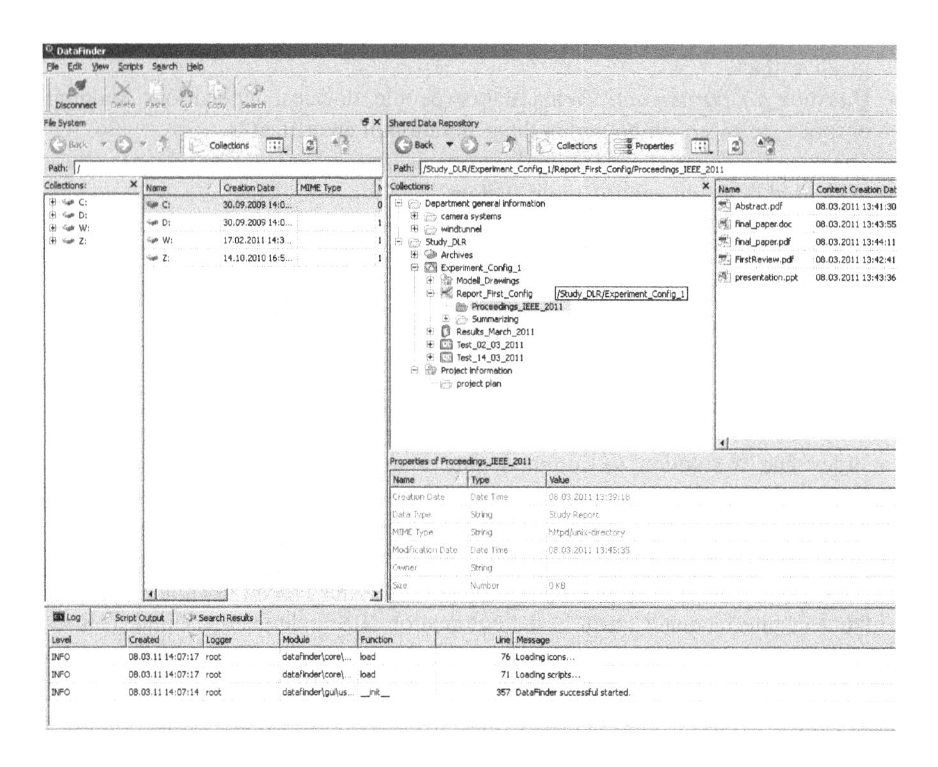

Fig. 7.1 User interface of the DataFinder.

It is designed similar to a file manager on common operating systems. The left hand side presents the local file hierarchy, and the right displays the shared repository. All data on the server can be augmented with arbitrary meta-data. Common actions available for both sides are: open, copy, paste, import and export data. Opening an entry will make an attempt to use the local system's default association for a file. These operations are all essential due to the nature of DataFinder being a data management tool. One must be aware that on some operations (e. g. copying) provenance related information is not copied with it. Copying would create a fork in the provenance graph to create a duplicate of a formerly uniquely referenced artifact. Special treatment to treat these cases in a way as to extend the graph properly are not in place, yet.

An advantage of DataFinder is, that an individual data model is configured for a shared repository, which must be followed by all its users. A data model defines the structure of collections. Collections can contain (configurable) allowed data types, that can be inserted into the collection. The data model also defines a pre-defined meta-data structure for these collections. This meta-data can be specified to be either optional or mandatory information when importing a data item. Based on the data model, data can be managed on a heterogeneous storage system (certain data items stored in different storage sub-systems, see 7.1.2.2). This requires that DataFinder provides the ability to manage data on different storage systems, under the control of a single user interface under a single view (even within the same collection).

Lastly, it is the possibility to extend the application with Python scripts. This enables a user to take advantage of more customised features, such as tool integration, task automation, etc.

The DataFinder-based system aims at providing many options and to be highly extensible for many purposes. DataFinder is already in use in different fields of research. New use cases are identified and extensions implemented frequently. One of these is the new use case for supporting a good laboratory practice capable notebook as outlined in this chapter.

7.1.2.2 Distributed Data Storage

One of the key features of DataFinder is the capability to use different distributed, heterogeneous storage systems (concurrently). A user has the freedom to store data on different systems, while meta-data for this data can be kept either on the same or on a different storage system.

Possible data storage options can be accessed for example through: WebDAV, Subversion, FTP, GridFTP. Other available storage systems possibilities are Amazon S3 Cloud services as well as a variety of hosted file systems. Meta-data for systems not capable of providing extensive free-form meta-data is managed centrally with another system. Such systems then are accessed through meta-data capable protocols like WebDAV or Subversion. Further storage back-ends are relatively simple to integrate, due to the highly modular factory design of the application. This design feature of DataFinder will be further examined in Sect. 7.3 for the integration of a distributed Grid data storage infrastructure.

It must be noted at this point however, that DataFinder is responsible for maintaining consistently managed data. DataFinder uses these protocols and systems for this purpose. If data is accessed *without* using the DataFinder directly on the server through other clients, data policies may be compromised (due to different access restrictions), or consistency may be compromised (with writing access to the storage systems). With certain caution, this can however be used to integrate other (legacy) systems into the overall concept.

Due to the design of the DataFinder it is further possible to manage physical (real world) items, such as laboratory analysis samples or offline media (e. g. video tapes, CDs, DVDs). Physical items can be stored on shelves, or archived in any other way. These can be valuable artifacts for research, and the knowledge of their existence as well as their proper management is a common necessity. Therefore, it is crucial to managed them electronically in a similar fashion by the same management tools. Doing so enables extensive meta-queries provided by the DataFinder, taking advantage of utilising the search capabilities over all managed items in the same way. Furthermore, this enables to reference them consistently in provenance assertions from within the realm of the provenance enabled system.

7.1.3 Overview

This chapter ties the link between the existing DataFinder application to convert it into a tool useful for a good laboratory practice compliant electronic notebook. It will introduce how DataFinder can be combined with provenance recording services and (Grid) storage servers to form the back bone of such a system. The concept of DataFinder is to be a system that can be customised towards different deployment scenarios, it is to support the researchers or engineers in *their* way of working. This includes the definition of a data storage hierarchy, required meta-data for storage items and much more, usually alongside with customisation or automation scripts and customised GUI dialogues. In a similar fashion, DataFinder can be used to construct an electronic laboratory notebook with provenance recording for good laboratory practice. Again, to do so one creates the required data models and customises GUI dialogues to suit the purpose.

The used provenance technologies and their applications are described in Sect. 7.2. Concepts to integrate Grid technologies for scientific data management are outlined in Sect. 7.3. Sect. 7.4 presents the results of integrating the good laboratory practice into a provenance system as well as a data management system. It also provides a solution on how to connect these two system practically. Finally, the concept of the resulting system of an electronic laboratory notebook is evaluated.

7.2 Provenance Management

Provenance originates from the Latin word: "provenire" meaning "to come from" [11]. It is described as "the place that sth. originally came from" thus the origin or source of something (cf. [22]). It was originally used for art, but other disciplines adapted it for their objects, such as fossils or documents. In the field of computer science and data origin it could be defined as:

> "The provenance of a piece of data is the process that led to that piece of data." [12]

Based on this understanding, approaches for identifying provenance use cases for modeling processes and for integrating provenance tracking into applications are developed. Also, concepts to store and visualise provenance information are investigated. An overview of the different areas of provenance gives Fig. 7.2.

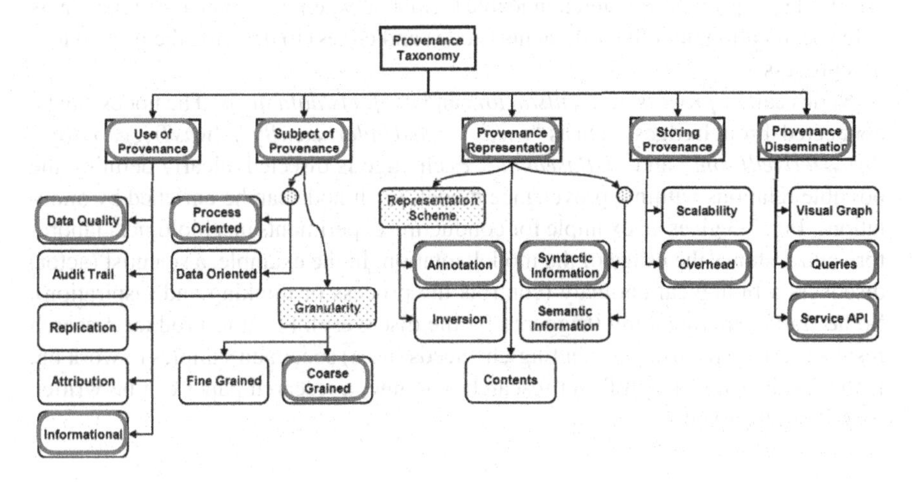

Fig. 7.2 Provenance taxonomy according to [19].

The figure shows five major areas: *Usage, Subject, Representation, Storage and Dissemination.* [19] gives a detailed description on each area and their subdivisions. In this application, embedded provenance tracking in the data management system enables DataFinder to provide information about the chain of steps or events leading to a data item as it is. The following list outlines relevant elements of the taxonomy from Fig. 7.2 (additionally framed elements):

Use of provenance: Provenance is used to present *information* of the origin of the data, but also to provide *data quality.*

Subject of Provenance: The subject is the *process* of conducting a study or experiment. It is focused on documentation. To identify the subject further, the Provenance Incorporating Methodology (PrIMe, Sect. 7.4.1) is used.

Provenance Representation: Provenance information will be represented in an *annotational* model, based on the Open Provenance Model (OPM, Sect. 7.2.1) and it will mainly hold *syntactic information.*

Storing Provenance: Provenance information will be stored in the prOOst (Sect. 7.2.2) system (can also hold additional information).

Provenance Dissemination: To extract provenance information, the provenance system can be queried using a graph traversal language (Sect. 7.2.2.2).

The main concepts of OPM and the provenance system prOOst are described in the following sections, whereas PrIMe is discussed in the scope of applying the technical system to the domain of good laboratory practice in Sect. 7.4.

7.2.1 OPM – Open Provenance Model

The Open Provenance Model [13] is the result of the third "provenance challenge" efforts [18] to provide an interchangeable format between provenance systems. In its core specification, it defines elements (nodes and edges) to describe the provenance of a process.

Nodes can be *processes, agents/actors* and *artifacts/data items.* The nodes can be connected through edges, such as *"used", "wasUndertakenBy", "wasTriggeredBy", "wasDerivedFrom"* and *"isBasedOn"*. Each edge is directed, clearly defining the possible relations within a provenance model. Each node can be enriched by annotations. Fig. 7.3 gives an example for conducting experiments in a biological laboratory and it shows the usage of the model notation. In the example, a scientist (actor) discovers a biological anomaly (controls the process of thinking and inspiration). So he starts experimenting (triggered by the discovery). For it to produce research results (derived from experimenting), he needs (uses) specimen samples to work on. If the results show a significant research outcome, a research paper can be written (based on) the results.

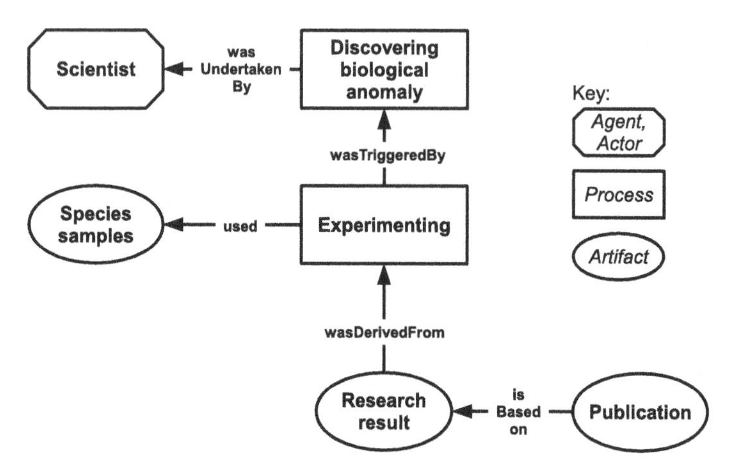

Fig. 7.3 Example a biological study as an OPM model.

7.2.2 Provenance Storage with prOOst

Groth et al. describe in [6] theoretically the architecture of a provenance system. In [14] the representation of a provenance system is described as follows: A provenance aware application sends information of interest to the provenance store. From this store inquiries and information is gathered, and possibly given back to the application.

To record the information, different approaches have been investigated. In [7] four different realisations are discussed: Relational, XML with XPath, RDF with SPARQL and semi-structured approaches. They conclude semi-structured approaches to be most promising. In semi-structured systems, the used technology has no formal structure, but it provides means of being queried.

This work uses a semi-structured approach for the provenance storage system *prOOst*. It uses the graph database "Neo4j" [3] for storage and the graph traversal language "Gremlin" [1] for querying. Furthermore, it provides a REST interface to record data into the store, and a web front end to query the database. The prOOst provenance system was published under the Apache license in July 2011 on SourceForge.[1]

It is not the first implementation using a graph database for storage technology. In [21] this approach was already successfully tested. Neo4j was chosen as it is a robust, performant and popular choice for graph storage systems. Additionally it readily connectible with the suitable Gremlin query system to meet our requirements. Further discussions on alternative storage or query systems are outside the scope of this chapter. Further information on the implementation of OPM model provenance assertions using these systems are described in the following two sections.

7.2.2.1 Graph Database: Neo4j

> "Neo4j is a graph database, a fully transactional database that stores data structured as graphs." (cf. [3])

An advantage of graph databases like Neo4j is that they offer very flexible storage models, allowing for a rapid development. Neo4j is dually licensed (AGPLv3 open source and commercial).

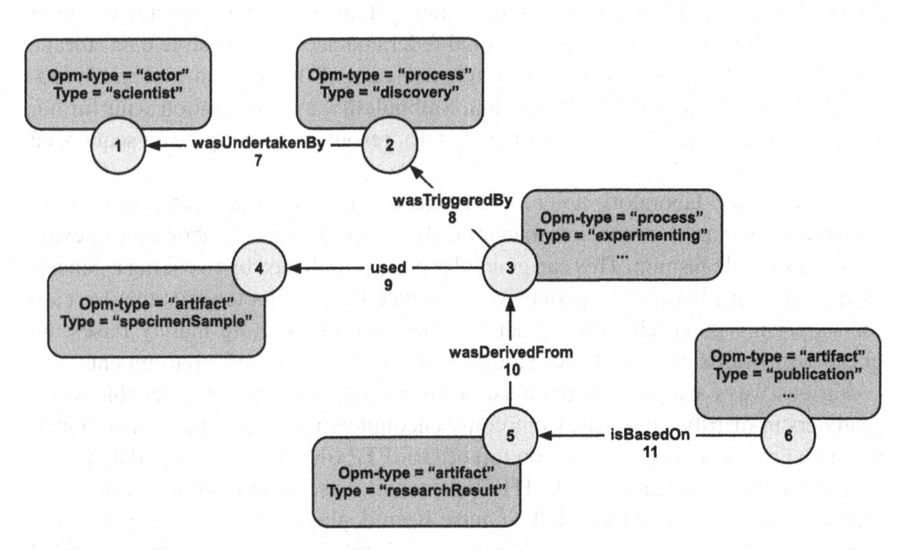

Fig. 7.4 OPM example in Neo4j.

[1] `http://sourceforge.net/projects/proost/`

Modelling OPM using Neo4j is described in more detail in [23]. Fig. 7.4 shows the previous example (from Fig. 7.3) modelled as an OPM graph. Each element is represented by a node (vertex) in the database. Nodes are indexed according to the Neo4j standard. The nodes can be annotated with further (OPM specific) information, such as "process" or "artifact". Analogously, also the edges connecting the nodes are indexed and annotated with a label (the OPM relationship).

7.2.2.2 Query Language: Gremlin

"Gremlin is a graph traversal language" [1]. Gremlin already provides an interface to interact with the Neo4j graph database. The following example shows its use for querying Neo4j on the example database, searching for the names (identifiers) of all discoveries of a certain `scientistX`:

```
$_g := neo4j:open('database')
$scientists := g:key($_g, 'type', 'scientist')
$scientistX := g:key($scientists, 'identifier', 'scientistX')
$discoveries := $scientistX/inE/inV[@identifier']
```

7.3 Distributed, Scientific Data Management

The previous sections have discussed the technical means to manage data on the user side (DataFinder) and to store and query the provenance information. As indicated, DataFinder can handle a variety of different data storage servers. However, to store data and its associated meta-data on the same system, and to take full advantage of Grid technologies for cross-organisational federated access, a suitable data storage service has to be chosen. For the example use case of the team of biologists, federated access management (e. g. through Shibboleth[2]) and integration with further Grid-based resources would be desired (e. g. for resources to compute on sequenced genome data).

An electronic laboratory notebook system is a data management system, only with the particular needs towards managing the experimental and laboratory relevant data in a suitable fashion. This can generally be accomplished by tweaking a generic storage system for data and (extensive) associated meta-data towards the use case for supporting good laboratory practice. This section therefore mainly raises the questions towards the use of such storage systems in Grid-based environments.

Various ways are possible to envision for making relevant data available to researchers in distributed teams. Commonly encountered mechanisms in such (Grid) research environments are based on top of GridFTP (the "classic" Grid data protocol) or WebDAV (extension to the HTTP protocol). In some environments more full featured infrastructures, like iRODS[3] have been deployed. One such environment is the New Zealand based "Data Fabric" – as implemented for the New Zealand eScience Infrastructure (during the recently concluded BeSTGRID project). iRODS

[2] `http://shibboleth.internet2.edu/`

[3] `http://www.irods.org/`

offers data replication over multiple geographically distributed storage locations, with one centralised meta-data catalogue. Its data is exposed through the iRODS native tools and libraries, as well as through WebDAV (using Davis[4]), a web-based front-end and GridFTP (through the Griffin GridFTP server [24, 25] with an iRODS back-end using Jargon[5]).

7.3.1 Integration of Existing Storage Servers

We are discussing data integration solutions according to the above mentioned scenario of the New Zealand Data Fabric. From this, slight variations of the setup can be easily extrapolated.

Three obvious possibilities exist to use this type of infrastructure for provenance enabled data management and/or as a laboratory notebook system for distributed environments. For all these, users need to be managed and mapped between multiple systems, as iRODS introduces its own mandatory user management. This may only be required for the storage layer, but it does introduce a redundancy. The options are discussed in the following paragraphs.

The easiest, and directly usable, way is to integrate this Grid Data Fabric as an external *WebDAV* data store, using the existing persistence module. Even though WebDAV is a comprehensive storage solution for the DataFinder for data and meta-data, this service layer on top of iRODS does not permit the required WebDAV protocol means to access the meta-data. An additional meta-data server is required, and therefore potentially multiple incompatible and separate sets meta-data may exist for the same data item stored. Unfortunately this WebDAV service does not use the full common Grid credentials for access, but is limited to MyProxy[6] based authentication as a work around.

As the next step up, DataFinder can be equipped with a *GridFTP* back-end in its persistence layer. Such a module was already available for a previous version (1.3) of DataFinder, and only requires some porting effort for the current (2.x) series. Again, GridFTP is only able to access the payload data, and is not capable to access any relevant meta-data, resulting in the need of an additional and separate meta-data service. An advantage is that this solution uses the common Grid credentials for authentication.

Lastly, the development of a native *iRODS* storage back-end based on the txIRODS Python bindings[7] is a possibility. This solution could also use the iRODS meta-data capabilities for native storage on top of the payload data storage. Unfortunately, this last solution also requires the use of the native iRODS user credentials

---

[4] WebDAV-iRODS/SRB gateway: `http://projects.arcs.org.au/trac/davis/`

[5] `https://www.irods.org/index.php/Jargon`

[6] Software for managing X.509 Public Key Infrastructure (PKI) security credentials: `http://grid.ncsa.illinois.edu/myproxy/`

[7] `http://code.arcs.org.au/gitorious/txirods`

for accessing the repository, as it is completely incompatible to any of the common Grid authentication procedures.

The above mentioned scenarios can be freely modified, particularly the first two regarding their underlying storage infrastructure. One could deploy other storage systems that expose access using WebDAV or GridFTP as service front ends for simplicity, potentially sacrificing any of the other desired features of iRODS like cross-site replication.

When sketching out a potential deployment, the above mentioned scenarios did not strike us as being particularly nice to implement or manage. Several shortcomings were quite obvious. Firstly, the central meta-data catalogue, which can turn out to be a bottle neck. Particularly meta-data heavy scenarios requiring extensive queries on meta-data would suffer due to increased latencies. Secondly, the iRODS system provides a multitude of features, which make the system implementation as well as its deployment at times quite convoluted. A simpler, more straight forward system is often preferred. Lastly, multiple user management systems can be an issue, particularly if this includes the burden of mapping between, particularly if they are based on different concepts. Grid user management is conceptually based on cross-organisational federation, including virtual organisations (VOs) and delegation using proxy certificates, which cannot be neatly projected to other user concepts as employed by iRODS.

7.3.2 Designing an Alternative Storage Concept – MataNui

The idea for an alternative storage solution came up, which is simpler and a better "Grid citizen." For performant storage of many or large files inclusive meta-data, the NoSQL database MongoDB with its driver side file system implementation "GridFS" seemed like a good choice. A big advantage of this storage concept is, that MongoDB can perform sharding (horizontal partitioning) and replication (decentralised storage with cross-site synchronisation) "out of the box." Therefore, the only concerns to target were to provide suitable service front-ends to the storage sub-system, to offer the capabilities for the required protocols and interfaces to the DataFinder. This means that research teams can opt for running local server instances (alternatively to accessing a remote server) for an increase in performance as well as decrease in latencies. This local storage sub-system also increases data storage redundancy, which leads to a better fault protection in cases of server or networking problems. Each storage server individually can be exposed through different service front-ends, reducing bottle necks. These service front-ends can be deployed in a site specific manner, reducing the number of server instances to those required for a site.

This distributed storage concept for data and meta-data, complimented by individual front-end services in a building block fashion, has been dubbed "MataNui" [10]. The MataNui server [9] itself provides full access to all content, including server side query capability and protection through native Grid (proxy) certificate authentication (X.509 certificates). As the authentication is based on native Grid means, it is obvious to base the user management on Grid identities as well, the distinguished names (DN) of the users. MataNui is based on a REST principle based Web Service (using JSON encoding), and is therefore easy to access through client side implementations.

Exposing further server side protocols is done by deploying generic servers, that have been equipped with a storage back-end accessing the MataNui data structures hosted in the MongoDB/GridFS containers. It was relatively simple to implement the GridFTP protocol server on the basis of the free and open Griffin [24, 25] server. A first beta development level GridFS back-end is already part of the Griffin code base. Possibly later a WebDAV front end is going to be implemented, equipping one of the quite full featured Catacomb[8] or LimeStone[9] servers with a GridFS back-end for data and meta-data. Such servers then could also be used to access (and query) the meta-data through the WebDAV protocol, if the storage back-end supports this. Lastly, it is even possible to use a file system driver to mount a remote GridFS into the local Linux/UNIX system. However, access control to the content is provided through the services on top of the MongoDB/GridFS server. Therefore, this will likely circumvent any protective mechanisms. A better solution would be to mount a WebDAV exposed service into a local machine's file system hierarchy.

Access through protocols as GridFTP and WebDAV is quite straight forward through various existing clients in day-to-day use within the eResearch communities. This is different with the MataNui RESTful service. As outlined in Sect. 7.3.1 already, the DataFinder can be quite easily extended towards providing further persistence back-ends, like a potential iRODS back-end. In a similar fashion a MataNui REST service client back-end can be implemented. The big difference being, that it does not require any external modules that are not well maintained. It can mainly be based on the already available standard library for HTTP(S) server access, with the addition of suitable cryptography provider for extended X.509 certificate management. This can be done either by simply wrapping the OpenSSL command line tool or by using one of the mature and well maintained libraries such as pyOpenSSL.[10]

This modularity of service front-ends leaves administrators the option to set up sites with exactly the features required locally. However, in a global perspective, MataNui enables a new perspective on the functionality of a data fabric for eResearch. Fig. 7.5 provides a conceptual overview of how such a distributed data repository can be structured. Every storage site requires an instance of MongoDB with GridFS. These are linked with each other into a replication set (with optional sharding). The storage servers for the different sites expose the repository through one or more locally hosted services, such as the MataNui RESTful Web Service, a GridFTP server, etc. These services can be accessed by clients suitably equipped for the particular service. Clients, such as the DataFinder, may require an additional implementation for a particular persistence back-end. Some of these clients (e. g. DataFinder or a WebDAV client) may be equipped to take advantage of the full meta-data capabilities of the data fabric, whereas others (e. g. GridFTP or file system

[8] `http://catacomb.tigris.org/`
[9] `https://github.com/tolsen/limestone`
[10] `https://launchpad.net/pyopenssl`

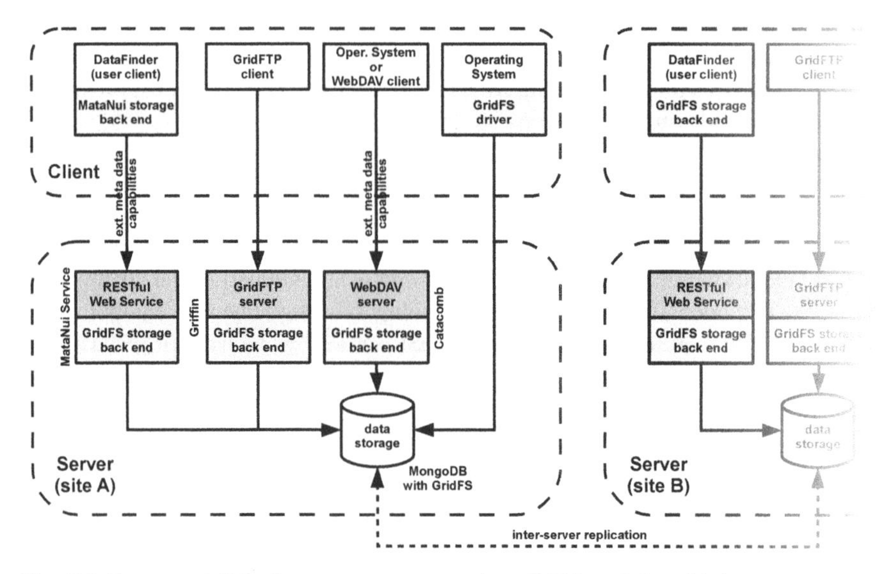

Fig. 7.5 Conceptual links between components in a Grid-based data fabric to support researchers in distributed environments. The system provides for decentralised access to geographically distributed data repositories, while enabling administrators to only expose local storage through service front-ends as required.

mounted WebDAV) may only access the data content along with some rudimentary system meta-data (time stamp, size, etc.).

In a scenario like this data and its meta-data can be managed in the distributed environment through DataFinder. Seamless integration when working with other Grid resources is unproblematic: All systems share the same type of credentials, and data can be transferred between Grid systems directly through GridFTP without the need of being routed through the user's workstation.

7.4 Results

The following describes the application of the previously discussed technologies to implement the provenance enabled electronic laboratory notebook. For this also the data management system DataFinder requires customisation (through Python script extensions) to suit the users' needs. It is enhanced with features to trace documentation.

First the development of the provenance model for good laboratory practice by means of the PrIMe methodology is described in Sect. 7.4.1. Required modifications applied in the DataFinder code are outlined in Sect. 7.4.2. Sect. 7.4.3 evaluates the integration of DataFinder for the purpose of use as an electronic laboratory notebook in a final system. More information on this evaluation can be found in [16]. Lastly, Sect. 7.4.4 gives an outlook on improving DataFinder in its role as an electronic laboratory notebook, as well as on deploying such an infrastructure fully to Grid environments.

7.4.1 Developing a Provenance Model for Good Laboratory Practice

Munroe et al. [15] developed the PrIMe methodology to identify parameters for "provenance enabling" applications. These parameters then can be used to answer provenance questions. A provenance question usually identifies a scenario, in which provenance information is needed. Questions relevant for the analysis, are for example: "Who inserted data item X?", "What data items belong to a report X?" and " What is the logical successor of data item X?".

This approach was modified (in [23]), as it used the older p-assertion protocol (p. 15 in [23] and p. 2 in [15]) instead of the now more common Open Provenance Model (OPM) [13]. The p-assertion protocol is similar in use to OPM, so the approach can easily be adapted. The following list describes the three phases of the adapted PrIMe version in correspondence to the PrIMe structure from Fig. 7.6:

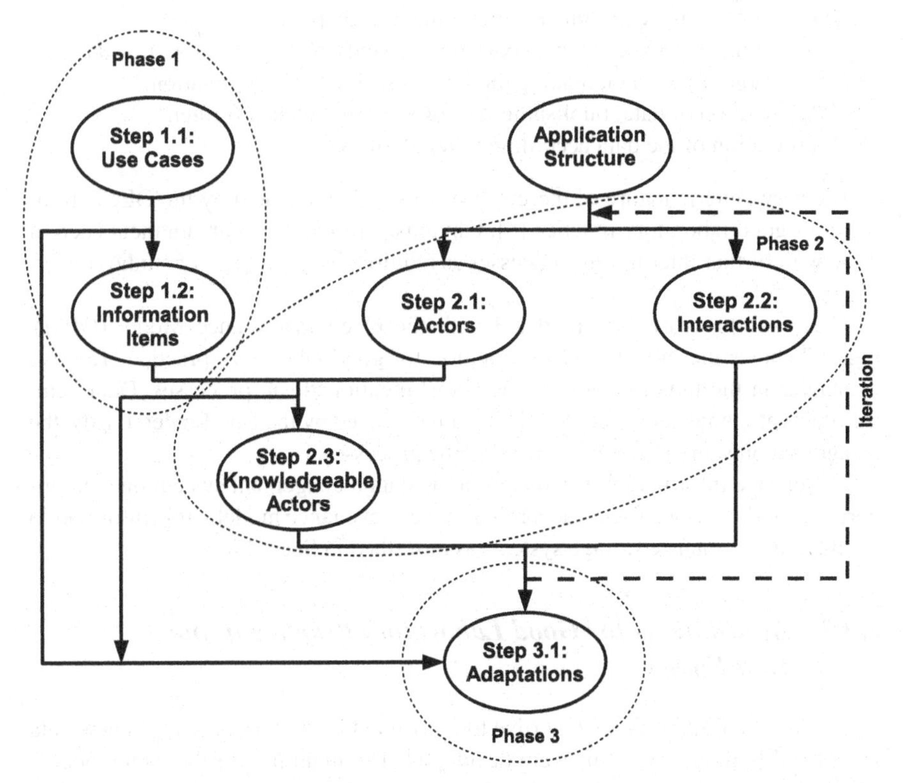

Fig. 7.6 Structure of PrIMe approach [15].

Phase 1: "In phase 1 of PrIMe, the kinds of provenance related questions to be answered about the application must be identified" [15] (p. 7). First, provenance *questions* are determined. Then, corresponding *data items/artifacts* that are relevant to the the answer, are investigated.

Phase 2: *Sub-processes, actors* and *interactions* are identified in phase 2. The sub-processes are part of the adaptation (Step 2.1). Actors generate data items or control the process. Relations between sub-processes and data items are defined as interactions (Step 2.2). Actors, processes and interactions are modeled with OPM.

Phase 3: The last phase finally adapts a system to the provenance model. In this phase, the provenance store is populated with information from the application. In the discussed scenario, this is accomplished via REST requests to the storing system.

Some exemplary questions that could be relevant in the sample use case have already been given in Sect. 7.1.1. After analysing the questions, participating processes need to be identified. A scientific experiment for which documentation is provided can be divided into five sub-processes:

1. Preparation of the experiment, generating a study plan.
2. Execution of the experiment according to a study plan, generating raw data.
3. Evaluation of raw data, making them processable for interpretation.
4. Interpretation of data, publishing it or processing the data further.
5. Preservation of the data according to regularities.

The very generic nature of these sub-processes is mandated by the OECD principles of good laboratory practice [8]. Obviously, researchers can augment each of these with further internal sub-processes as required by the project or studies undertaken.

These sub-processes are modeled with the Open Provenance Model (OPM). Fig. 7.7 shows the model in OPM notation for good laboratory practice. The five rectangles in the figure symbolise the above mentioned sub-processes. Data items/artifacts are indicated by circles: These are managed by the DataFinder. Lastly, the octagons represent the actors controlling the processes.

Provenance information is gathered in the data management system on data import and modification. Then – according to the provenance model – this information is sent to a provenance storage system (as described in Sect. 7.2).

7.4.2 Adjustments for Good Laboratory Practice in the DataFinder

To use the DataFinder as a supportive tool for good laboratory practice, a new data model and Python extensions were developed. The main part of the data model is presented in Fig. 7.8.

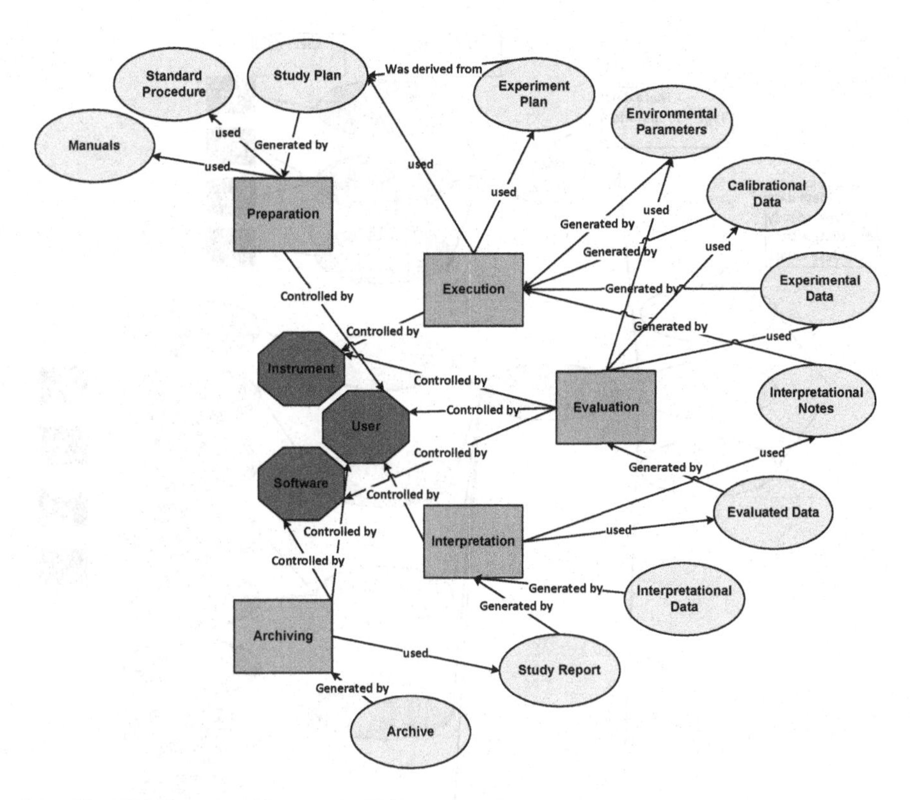

Fig. 7.7 OPM for scientific experiment documentation.

The model is derived through requirements analysis in [8]. It divides the data into the five major categories according to Sect. 7.4.1: Preparation, execution, evaluation, interpretation and archiving. All experiments pass through theses categories in their five processes. Each process needs or generates different types of data. Data is aggregated in (nested) *collections,* the data repository equivalent of directories in a file system. Collections representing these processes aggregate data items belonging to that process. Each collection or element can mandate attached meta-data (such as type or dates). The data model also provides structural elements at a higher level of the hierarchy to differentiate between different studies and experiments. Processes and data items are reflecting the model structure in Fig. 7.7. The DataFinder repository structure is defined through its underlying data model[11] (implemented according to the OPM model).

In the screen shot of Fig. 7.1 at the beginning of this chapter, a user is connected to a shared repository (left side) operating on the described data model. The user now is required to organise data accumulated according to this model. For example,

[11] The complete data model is described in XML and available on `https://wiki.sistec.dlr.de/ DataFinderOpenSource/ LaboratoryNotebook`

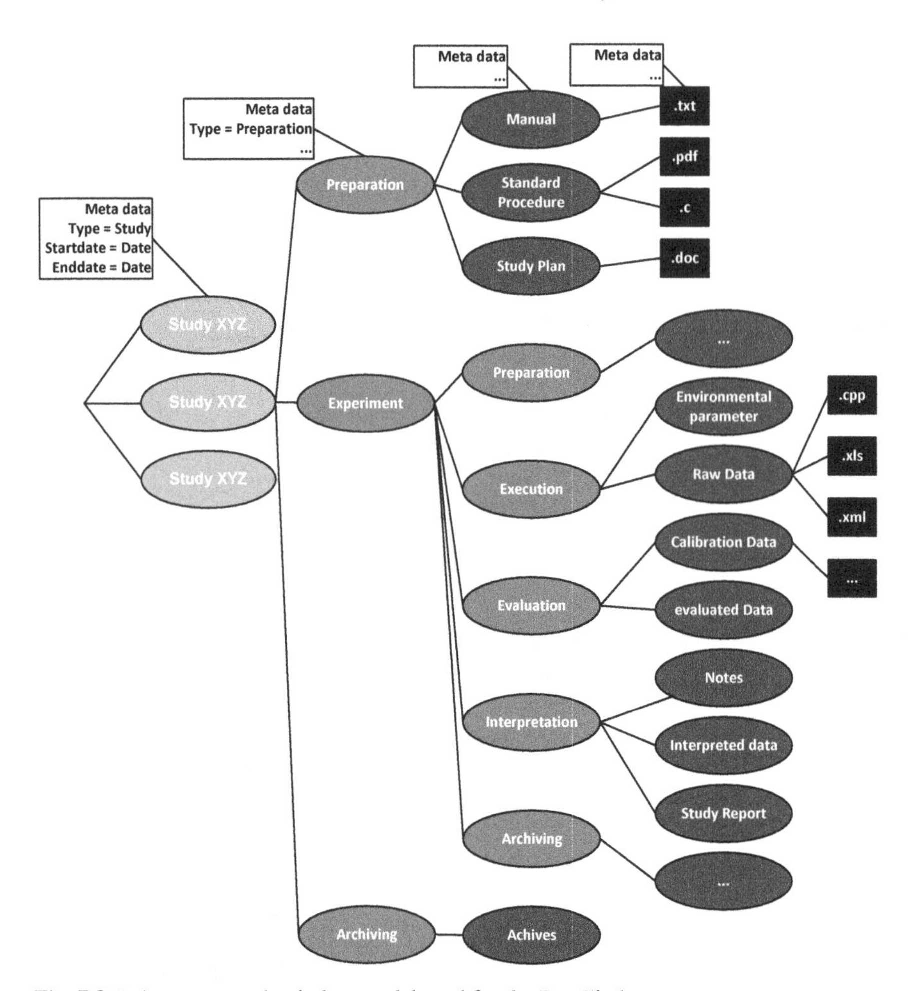

Fig. 7.8 Laboratory notebook data model used for the DataFinder.

a new collection of manuals may only be created within a parent collection of the "preparation" type. A "preparation" collection can then be either part of a "study" or "experiment."

Three further extensions to DataFinder have been developed. They are needed to support good laboratory practice in DataFinder:

- The most important extension is an *observer mechanism,* listening on *import events* into the DataFinder. Upon the import of a new document, it reacts by prompting with a dialog asking for input items within the system that have influenced the data item/artifact. After analysing the corresponding process, the information is recorded in the provenance store.

- A second extension supports *evidential archiving*. For this the user can send an archive to an archiving service, to analyses the credibility of the archive.[12] To provide sufficient information, the user can activate a specific script extension, which generates an archive composed of information relevant to the data from information in the provenance store. The user selects a study report, and the provenance store is queried for all data items influencing the report for each step.
- Lastly, a *digital signing mechanism* was implemented through an extension, aiming at increasing credibility of data items through non-repudiation.

7.4.3 Integration Evaluation of an Electronic Laboratory Notebook

Tab. 7.1 evaluates the DataFinder concepts on the requirements defined in Chap. 3.1 of [16]. It explains how each requirement is integrated into the DataFinder system.[13] The table shows that almost all requirements are either already currently met, are implemented through extensions as described here, or otherwise currently implemented. As a result, DataFinder can be used as laboratory notebook, supporting the concepts of good laboratory practice, and is therefore supportive to scientific working methods.

7.4.4 Outlook: Improving DataFinder-Based Laboratory Notebook

Of course the systems discussed in this chapter themselves are still research in progress and under constant development. On the one hand we can already envision a list of laboratory notebook features for desired improvements. On the other hand, this DataFinder-based laboratory notebook can be deployed to Grid environments.

7.4.4.1 Improving Laboratory Notebook Features

After the implementation, the next step is to deploy and integrate the system not only as a data management system, but as laboratory notebook to suit the needs of different organisations. For every deployment, customisation through automation scripts and specialised GUI dialogues need to be performed. Particularly for the purpose of electronic laboratory notebooks, some generic and easy integrated note editor widget would be much appreciable for free form note taking (instead of using external editors and importing the resulting data files). One a much more specialised level the following future features are considered to be beneficial to further improve the laboratory notebook functionality of DataFinder, and therefore meet the requirements of other deployment scenarios:

[12] This is not further discussed, because it is a separate project in Germany.

[13] The table and its description is adapted from [16]

Table 7.1 Implementation of the laboratory notebook requirements into the DataFinder

Requirement	Implemented?	Details
• Chain of events	yes with extensions described here	provenance for modeling the use case and storing the information
• Durability	yes	with extension from this application, but also through former solutions
• Immediate documentation	under development	a web portal is implemented
• Genuineness	yes customisation issue	combination of work flow integration in the DataFinder and the provenance service
• Protocol style	yes original	can be added as files to the system
• Short notes	yes original	as extra files or meta data to a data item
• Verifying results	yes (rudimentary)	signing concept and implementation as extension
• Accessibility	yes original	open source software
• Collaboration	yes original	same shared repository for each user, with similar information
• Device integration	yes customisation issue	integration via script API
• Enabling environmental specialisation	yes customisation issue	can be customised with scripts and data model
• Flexible Infrastructure	yes original	client: platform independent Python application server; meta data: WebDAV or SVN (extensible); data: several (extensible)
• Individual Sorting	partly under development	customising the view of the repositories is possible (but saving the settings is in planning)
• Rights/privilege management	yes under construction	the server supports it on the client side, the integration into DataFinder is currently developed
• Variety of data formats	yes original	any data format can be integrated, opening them depends on the users system
• Searchability	yes original	full text and meta data search
• Versioning	yes	SVN as storage back-end is developed to enable versioned meta data and data

Mobile version of DataFinder: A mobile version of the data management system client could ease the scientist's documentation efforts when working on-site, away from the established (office, lab) environment. This way the scientist could augment data items through notes or add/edit meta-data and data on-the-fly. Requirement of immediate documentation could be met through this extension.

Automatic generation of reports: For many project leaders it is interesting or important to be kept up-to-date on the current status of a project or what their team members are currently doing. For this they can currently only access the data directly. A feature summarising current reports and gives an intermediate report, could simplify the check up. This feature was found in the evaluated laboratory notebook mbllab [2].

Integrated standard procedures: In GLP, a standard procedures defines workflows for specific machines. In the laboratory notebook mbllab [2] these are integrated and give the user a guideline for actions.

More elaborate signing and documenting features: Scientists discuss results of colleagues. For more collaborative work situations, DataFinder needs to be enhanced with better features for user interactions. On the one hand a discussion/-commenting mechanism on data items could be supported, on the other hand a scientist can sign data and leave some kind of digital identity card. This could be used to reference a list of other items signed or projects worked on. In the evaluated laboratory software NoteBookMaker [4], a witness principle with library card is integrated. Each notebook page contains an area, where a scientist can witness (authenticate) an entry. After witnessing the data, the information of the witnessing person's identity is displayed on the corresponding page. This witnessing information is then connected to a library card listing personal information and projects.

A graphical representation: A graphical representation of the provenance information on the server or in the DataFinder can help to make provenance information visually more accessible. This integration of provenance data in DataFinder assists a user in understanding correlations between items.

Configuration options: Selecting a specific provenance or archiving system should be possible. This could be handled through a new option in the data store's configuration. Additionally a dialog prompting for this information needs to be implemented.

7.4.4.2 Migrating the Laboratory Notebook to the Grid

Sect. 7.3.2 already explains how a data management system suitable for the Grid can be constructed. The laboratory notebook system is "resting" on top of that particular data storage system, under support of a provenance store to enable provenance enabled working schemes. Therefore, the two aspects of an underlying Grid-based data storage system and of a Grid-enabled provenance store need to be discussed.

While MongoDB with GridFS is a mature product ready to deploy, the overlaying service infrastructure for a Grid-enabled data service is not quite as matured. Currently the Griffin GridFTP server [25] is in productive deployment both in the Australian as well as the New Zealand eScience infrastructures. However the GridFS storage back-end already works, but is still only available in a beta version and needs a little further completion and testing. The situation is similar with the MataNui RESTful Web Service front end, which still needs implementation of further query functionality. Current tests of the two systems have showed that throughput bottle necks to both services currently seems mostly limited by the throughput of the underlying disk (RAID) storage system or network interface (giga-bit ethernet), while the database and service layer implementation is easily holding up even on a moderately equipped system (CPU and memory).

To access this MataNui infrastructure with the DataFinder at least one of two things still has to be implemented: The GridFTP data store back-end needs to be ported from the 1.x line of DataFinder versions, or a MataNui data store back-end needs to be implemented for the current version. For best results preferably the latter has got priority on the list of further implementations to reach this goal. Due to the nature of the service as well as the persistence abstraction in the 2.x DataFinder versions, this should be relatively straight forward. This enables DataFinder to completely retire WebDAV or Subversion as a centralised data server for data content as well as meta-data, relocating this information completely onto a Grid infrastructure.

In such a setup, DataFinder accesses the MataNui service natively, while all managed (payload) data can be accessed through GridFTP (Griffin server) for the purpose of compatibility with other Grid environments. This supports common usage for example using file staging for Grid job submission. Storage server side replication ensures seamless usage in geographically distributed research teams while retaining high throughput and low latencies through the geographically closest storage server.

The provenance store prOOst currently does not yet support access of its REST service through Grid authenticated means. Once this is implemented for the newly releases provenance store, every required service for a Grid-enabled data service with provenance capabilities, can be accessed using the same credentials and common Grid access protocols.

7.5 Conclusion

This chapter sketches a scenario of using provenance tracking with DataFinder to support good laboratory practice and to track relations between stored documents. In this scenario DataFinder is used in a distributed system together with a central provenance store. This makes it possible to access and update data from virtually anywhere with a network connection, while keeping track of all interactions with data items through recorded provenance information at any time. When implementing the laboratory notebook, stored provenance information can be queried to enable the extraction of additional valuable meta-data information on data items. As a result, provenance is successfully used to trace typical scientific workflows comprising of preparation, execution, evaluation, interpretation and archiving of research data. The reliability – and therefore credibility – of research results is increased, and assistance to help understand involved processes is provided for the researcher.

Such a system can be implemented on top of a Grid data infrastructure, as the described MataNui system. The MataNui service is mostly functioning already, but still needs integration into DataFinder as a full-featured storage back-end for data as well as meta-data. Additionally, it is already possible to expose the data repository to Grid environments directly using the GridFTP protocol. GridFTP is commonly used for scripts, automation and compatibility with other Grid enabled tools. The overall MataNui concept has been designed to be capable of handling files large in number and size, as well as manage arbitrary amounts of meta-data associated with each data item. It is usable in distributed projects with a self-replicating, federated data infrastructure. This federation can drastically improve data access latency and throughput by connecting to a geographically close service. Through support for server side queries, meta-data searches can be processed very efficiently by avoiding transfers of potentially large numbers of data sets to a client. Lastly, the implementation of MataNui has been undertaken with the vision of it being robust as well as easy to deploy and use.

References

1. Gremlin graph traversal language Web Site, https://github.com/tinkerpop/gremlin/wiki
2. mbllab–Das elektronische Laborbuch, http://elektronisches-laborbuch.de/
3. Neo4j Graph Database Web Site, http://neo4j.org/
4. Note Book Maker for PC and Mac, The World Leader in Virtual NoteBooks, http://www.notebookmaker.com
5. Buneman, P., Khanna, S., Tan, W.C.: Why and Where: A Characterization of Data Provenance. Tech. rep., University of Pennsylvania (2001), http://repository.upenn.edu/cis_papers/210/
6. Groth, P., Miles, S., Tan, V., Moreau, L.: Architecture for Provenance Systems (2005), http://eprints.ecs.soton.ac.uk/11310/
7. Holland, D.A., Braun, U., Maclean, D., Muniswamy-Reddy, K.K., Seltzer, M.I.: Choosing a Data Model and Query Language for Provenance. In: Proceedings of the 4th International Provenance and Annotation Workshop, IPAW (2008), doi:10.1.1.152.3820
8. Inter-Organization Programme for the Sound Management of Chemicals (IOMC): No 1: OECD Principles on Good Laboratory Practice (1998), http://www.oecd.org/document/63/0,2340,en_2649_34381_2346175_1_1_1_37465,00.html
9. Kloss, G.K.: MataNui Project, http://launchpad.net/matanui (last accessed June 2011)
10. Kloss, G.K.: MataNui – Building a Grid Data Infrastructure that "doesn't suck!". In: Proceedings of the 1st New Zealand eResearch Symposium, Auckland, New Zealand (2010)
11. Merriam Webster, I. (ed.): Merriam-Webster Online Dictionary. Merriam-Webster, Incorporated (2010)
12. Moreau, L.: The Foundations for Provenance on the Web. Foundations and Trends in Web Science 2(2-3), 99–241 (2010), http://eprints.ecs.soton.ac.uk/21691/

13. Moreau, L., Clifford, B., Freire, J., Futrelle, J., Gil, Y., Groth, P., Kwasnikowska, N., Miles, S., Missier, P., Myers, J., Plale, B., Simmhan, Y., Stephan, E., den Bussche, J.V.: The Open Provenance Model core specification (v1.1). Future Generation Computer Systems 27(6), 743–756 (2010), `http://openprovenance.org/`, doi:10.1016/j.future.2010.07.005
14. Moreau, L., Clifford, B., Freire, J., Gil, Y., Groth, P., Futrelle, J., Kwasnikowska, N., Miles, S., Missier, P., Myers, J., Simmhan, Y., Stephan, E., den Bussche, J.V.: The Open Provenance Model—Core Specification (v1.1). Future Generation Computer Systems 27, 743–756 (2010), `http://eprints.ecs.soton.ac.uk/21449/`, doi:10.1016/j.future.2010.07.005
15. Munroe, S., Miles, S., Groth, P., Jiang, S., Tan, V., Moreau, L., Ibbotson, J., Vazquez-Salceda, J.: PrIMe: A Methodology for Developing Provenance-Aware Applications. Tech. rep., Grid-Provenance Project, Southampton, UK (2006), `http://eprints.ecs.soton.ac.uk/13215/`
16. Ney, M.: Enabling a data management system to support the good laboratory practice. Master's thesis, Free University of Berlin (2011), `https://wiki.sistec.dlr.de/DataFinderOpenSource/LaboratoryNotebook`
17. Schlauch, T., Schreiber, A.: DataFinder – A Scientific Data Management Solution. In: Proceedings of Symposium for Ensuring Long-Term Preservation and Adding Value to Scientific and Technical Data 2007 (PV), Oberpfaffenhofen, Germany (2007)
18. Simmhan, Y., Groth, P., Moreau, L.: Special Section: The third provenance challenge on using the open provenance model for interoperability. Future Generation Computer Systems 27(6), 737–742 (2011), `http://www.sciencedirect.com/science/article/pii/S0167739X100%02402`, doi:10.1016/j.future.2010.11.020
19. Simmhan, Y.L., Plale, B., Gannon, D.: A Survey of Data Provenance Techniques. Tech. rep., Computer Science Department, Indiana University, Bloomington, IN, USA (2005), doi:10.1.1.70.6294
20. The Data Finder Team: DataFinder Project, `http://launchpad.net/datafinder` (last accessed June 2011)
21. Tylissanakis, G., Cotronis, Y.: Data Provenance and Reproducibility in Grid Based Scientific Workflows. In: Workshops at the Grid and Pervasive Computing Conference, pp. 42–49 (2009), doi:10.1109/GPC.2009.16
22. Wehmeier, S. (ed.): Oxford Advanced Learners Dictionary, 6th edn. Oxford University Press (2000)
23. Wendel, H.: Using Provenance to Trace Software Development Processes. Master's thesis, University of Bonn, Bonn, Germany (2010), `http://elib.dlr.de/64835/`
24. Zhang, S., Coddington, P., Wendelborn, A.: Connecting arbitrary data resources to the Grid. In: Proceedings of the 11th International Conference on Grid Computing (Grid 2010). ACM/IEEE, Brussels (2010)
25. Zhang, S., Kloss, G.K., Behnke, L.: Griffin Project (2011), `https://projects.arcs.org.au/trac/griffin` (last accessed March 2011)

Author Index